Optimal Thinning within the Faustmann Approach

Renke Coordes

Optimal Thinning within the Faustmann Approach

 Springer Vieweg

Renke Coordes
Tharandt, Germany

Also Doctoral Thesis, university Technische Universität Dresden, 2013

ISBN 978-3-658-06958-2   ISBN 978-3-658-06959-9 (eBook)
DOI 10.1007/978-3-658-06959-9

The Deutsche Nationalbibliothek lists this publication in the Deutsche Nationalbibliografie; detailed bibliographic data are available in the Internet at http://dnb.d-nb.de.

Library of Congress Control Number: 2014946616

Springer Vieweg
© Springer Fachmedien Wiesbaden 2014

Printed on acid-free paper

Springer Vieweg is a brand of Springer DE.
Springer DE is part of Springer Science+Business Media.
www.springer-vieweg.de

Foreword

The problem of optimal thinning is highly relevant for practical forestry. Thinning generates a substantial part of the net revenue, and it shapes the future value development of the remaining forest stand. Particularly with regard to the currently observed strong shifts of forest growth, timber prices and the compositions of the product classes, a profound theoretical foundation is necessary. Astonishingly, down to the present day, the problem of optimal thinning has not been solved satisfactorily. According to my judgment, the reason for this is the high degree of complexity of this optimization problem which is often shrouded.

In the dissertation, Renke Coordes has solved the problem of optimal thinning.

He succeeded in presenting a solution with an own methodical approach in which the optimum optimorum is not to be found necessarily. Also, Renke Coordes consequent applies the Faustmann approach and follows a strong economic logic. In the dissertation, Renke Coordes presents a closed economic investigation of thinning. It is the best what I have read in this research field during the last decades.

The dissertation is not only a fairly sizeable theoretical improvement; moreover, it has guiding character for practical silviculture. Questions on the optimal thinning age, the optimal thinning regime, the problem of quality and quantity, etc. are answered. In addition, well-known Central European principles of silviculture are tested for their theoretical foundation. Furthermore, the model as well as the analysis of the dissertation is of value in the broader field of natural resource economics.

I recommend the dissertation warmly for reading and studying to forest scientists as well as to forestry practitioners, and to natural resource economists, too.

Peter Deegen

Professor in the field of Forest Resource Economics,
University "Technische Universität Dresden", Germany

Table of Contents

List of Abbreviations and Symbols

Subscripts to variables indicate derivatives. Superscripts and subscripts to parameters indicate indices. Mathematical expressions are referred to in square brackets.

A	area	ha	hectare
a	level of financial assets	i	tree index number
$\hat{a}$	annual cost	$\|J\|$	Jacobi determinant
a_0	wealth endowment	j	tree index number
b	tree/stem shape reduction factor	k	tree index number
		L	Lagrange function
C	regeneration cost	l	stem length
CCD	cost-covering diameter	LV	land value
C_f	fixed regeneration cost	LEV	land expectation value
C_v	variable regeneration cost	LEV^a	analytical LEV of the Faustmann model
C_2	regeneration costs after a catastrophic event	LEV^b	LEV with only two harvest ages
c	consumption		
cc	cost-covering	LEV^F	Faustmann LEV
d	stem diameter of a tree	LEV^g	LEV with a constant rate of value change
$\hat{d}$	non-forestry income		
e	Euler's number	LEV^H	LEV with fixed harvest cost
F	Faustmann		
f	function	LEV^m	LEV with endogenous initial density
FPO	Faustmann-Pressler-Ohlin theorem		
		LEV^r	LEV with risk of catastrophic events
FV	forest value		
g	rate of value change	LEV^u	uneven-aged LEV
H	fixed harvest cost	LEV^1	exogenous thinning LEV
$\|H\|$	Hessian determinant	LEV^2	endogenous thinning LEV
h	variable harvest cost	LEV^β	solitary LEV
$\hat{h}$	tree height	LEV^+	intermingled LEV

m initial density

m^3 cubic meter

n number of trees

p net timber price or net unit timber revenue

$p.\,a.$ per annum

PIR Pressler's indicator rate

PV present value of land

Q timber volume of a stand

q timber volume of a tree

r continuous real market rate of interest

s stand age

si site index

s^T short-run timber supply

T Faustmann rotation age

t harvest age of a tree

t^* optimal harvest age of a tree

tc total cost-covering

$TCCD$ total cost-covering diameter

TIR Thinning indicator rate

ToC threshold of completion

ToM threshold of mortality

$TVCCD$ total variable cost-covering diameter

U intertemporal utility

u utility

V particular timber volume

v annual real market rate of interest

vc total variable cost-covering

V^n net value at the end of a period rotation

w annual discount factor

y number of stem

z vector of endogenous variables

α timber volume growth rate

β arbitrary factor

γ Poisson parameter

δ subjective rate of time preference

λ Lagrangian multiplier

ρ unit timber revenue

π pi

φ thinning-unaffected tree timber volume

σ additional tree timber volume

List of Figures

List of Tables

1 Problem

> "It is the great lesson which science has taught us that
> we must resort to the abstract where we cannot mas-
> ter the concrete. The preference for the concrete is to
> renounce the power which thought gives us."
>
> F. A. von Hayek (1964, p. 12)

Forest owners are observed to harvest trees in even-aged forest stands prior to the rotation period. These harvests without the intention of subsequent regeneration will be termed thinnings. Although numerous ecological and economic effects of thinnings have been discovered, the underlying motives of forest owners to thin forest stands have not been studied consistently. This is surprising since thinnings are long established and frequently conducted operations in forestry. From this perspective, thinnings are central elements of forestry which are directly interlinked with any other forestry operation. Regardless of the objectives, then, thinnings, if or when conducted, will influence their achievements crucially.

Doubtless, thinnings have been subject to forestry science from its very beginning (e.g. Carlowitz 1713, p. 201 ff.; Moser 1757, p. 221 ff.; Pfeiffer 1781, p. 143 ff.; Hartig 1791, p. 7 ff.; Cotta 1817, p. 42 ff.; Pfeil 1820, p. 294 ff.; Hundeshagen 1821, p. 103 ff.). Since then, countless effects of thinnings have been analyzed: on the growth of trees (e.g. Zhang et al. 2006), on other living creatures including humans (e.g. Bendz-Hellgren and Stenlid 1998; Rydberg and Falck 1998; Kalies et al. 2010; Neill and Puettmann 2013), on the soil (e.g. Tang et al. 2005), on the microclimate (e.g. Weng et al. 2007), on the water regime (e.g. Serengil et al. 2007), on carbon stocks (e.g. Garcia-Gonzalo et al. 2007), on the cash-flow (e.g. Cameron 2002), on the profitability (e.g. Hyytiäinen and Tahvonnen 2002), on the biodiversity (e.g. Wilson and Puettmann 2007), and on many others.

Due to the immense complexity, thinning effects are often analyzed separately. This analytical approach, however, prevents thinnings from being studied consistently as a human action. Certainly, potential advantages of thinnings come to mind quickly, but what about all the other, potentially disadvantageous effects? If thinnings, on the other hand, are beneficial without exceptions, why are some forest stands not thinned? From the large set of thinning effects, any effect might be used to seemingly explain any observation. If we infer that a forest owner thins a forest stand in order to improve the growth of some trees, we can argue in the same direction with reference to any other effect. If thinnings are addressed detached from the corresponding problem, the argumentation remains empty. Therefore, in order to derive empirically testable hypotheses, precisely defined conditions are necessary which restrict the set of possible explanations unambiguously (Popper 2002a, p. 36).

Without recognition of the patterns underlying our problems, it is impossible to distinguish between the countless things potentially observable (Hayek 1964). If we want to know why people sometimes harvest trees in exchange for other goods and services while retaining similar trees in other situations, why they sometimes do not supply timber for the development of a society, why they have incentives to harvest timber which is valuable to other people as standing trees (either alive or dead), why they sometimes prefer to harvest trees without considering to regenerate new trees or with the intention to convert the forest land to a different land use, we need to assess how forest owners act under specified circumstances in order to track those rules underlying the corresponding situation which induced the unexpected or unintended actions. On the basis of this explanation, solutions to problems can be outlined which cause to distort the consensus within a society.

The economic approach to these problems followed in this study is two-staged (cf. Homann and Suchanek 2005). First, the action of the individual is analyzed on the action level where individuals are assumed to follow the incentives of the situation. Individuals are the indivisible basic units of the analysis. As humans, they are assumed to be the only source of value. This "methodological individualism" (Hayek 1942) assigns objectives and actions to individuals exclusively. Incentives are valuated comparisons of individuals on the basis of expected utilities which define their subjective scale of preferences. By preferring actions promising higher utility to actions promising lower utility, individuals act as if maximizing utility. Finally, a situation is the totality of all constraints to the action of individuals as they perceive them. The sources of constraints are manifold. Technical constraints are given by the ecological possibilities to produce timber while individual constraints restrict the opportunities of the individual to achieve intended aims. Furthermore, on the action level, all actions of other individuals within the society are given as restrictions.

In the second step of the economic analysis, the interaction of different individuals is analyzed. On this interaction level, which constitutes the actual propositions for solutions to problems in a social world, action theory is applied to interactions such that the individuals can react to each other with the consequence of conflict or cooperation. In contrast to the action theory, the results of interaction theory are not ascribed to the objectives but to the conditions of the action. Instead, the realization of potential mutual gains from exchange is hampered by problems in the incentives and the information as each participant perceives them. It is these interaction structures of conflict and cooperation, i.e. of dilemma, that form the core of the economic analysis as understood in this study (Homann and Suchanek 2005).

On either level, individuals are assumed to be acting as Homo economici, i.e., as rational, self-centered utility maximizer. As a methodological principle (Popper 1983), the theoretical construct of the Homo economicus serves to focus the attention on the conditions of individual actions. It is thus neither a psychological theory about the behavior or thoughts of people nor it is a hypothesis about human action, as mistakenly postulated (cf. Kahneman 2011). Instead, as premises, actions are axiomatically rational by definition. The corresponding deductions (e.g. Varian 2010, p. 33 ff.) are thus tautological, circular and not testable since irrational actions are non-existent by assumption. Not until the inclusion of the conditions of the action, the economic hypothesis is formed, which is then empirically testable. These exogenously given constraints restrict the generality of the rationality assumption in order to separate a set of falsifiable propositions (Popper 2002b, p. 68 ff.). Although Homo economici might also be normatively justified (Buchanan and Brennan 2000, p. 74), the methodological employment in this work is purely positive.

The present study is a contribution to the action theory within forestry science. Hence, as exposed above, it serves as a foundation to actual interactive problems of conflict or cooperation. In this way, the derived propositions serve as hypotheses of individual action in dilemma situations, but not as predictions of isolated actions. The Faustmann model (cf. Chapter 3), as the most widely employed investment model in forestry science (Chang 2001), provides a distinct and consistent explanation for the phenomenon that people are observed to harvest and plant trees under precisely delimited conditions within the extended order of free market exchanges. Accordingly, prices as carrier of information provide forest owners with information about the valuation of goods and services of all other individuals (Hayek 2002). Through this process, forest owners receive incentives as to how they have to coordinate their actions in order to realize their personal objectives (Deegen et al. 2011). By harvesting trees, forest owners thus supply timber for exchange when it is valuable to other individuals. In this way, forest owners plant trees for the opportunity of future income. Whether these owners personally induce or conduct the harvest is irrelevant since they can equally sell the forested land which is valuable to other individuals due to the future harvest possibilities (Samuelson 1976).

Nevertheless, the Faustmann model provides no insights into the incentives which might induce forest owners to harvest trees prior to the rotation period. Instead, all trees share the same optimal harvest age per definition. It follows that the Faustmann model is unable to explain elementary operations within forestry. Thinnings are here not a minor detail. Instead, comprehension of their emergence might help clarifying the conditions, for instance, under which forest owners are induced to exploit forest stands without concern for subsequent regeneration. Or else, it might provide an explanation for the adjustment of the long-term investment in timber in an uncertain world of unanticipated changes or for the incentives which induce forest owners to supply more or less conservation services. Without a clear conception of how forest owners will respond to the conditions of a particular situation, it will be impossible to detect those reasons which lead to the observed deficiencies. Against this background, an extension of the "Faustmann laboratory" (Deegen et al. 2011, p. 363) to include thinnings appears highly desirable.

In order to provide an explanation for that part of human interactions which are based on voluntary and bilateral exchanges, the present study is based within the Faustmann approach. Thinnings are thus analyzed from the perspective of profitable timber production. The explored problem is demarcated by the generation of income with the production and sale of timber through the investment in tree growth. Since forest owners are assumed to act as rational Homo economici (see above), they act as if maximizing their intertemporal income with the production of timber as this offers the largest set of consumption opportunities (Hirshleifer 1970, p. 47f.). The larger set is preferred since it offers the same opportunity as the smaller set while, at the same time, providing additional opportunities. In a purely competitive market, on the other hand, maximization remains as the only feasible opportunity.

From the commitment to the Faustmann approach it follows that the arising propositions are restricted to the impersonal exchanges via markets within the extended order of human interaction (Deegen 2012). Only within this open society (Popper 1966), individuals respond to the abstract rules and impersonal signals generated by the "perfect" markets required for the analysis (cf. Section 3.3.1). On the other hand, the Faustmann model is a market model. Its underlying assumptions are the result of bilateral exchanges between individuals which aggregate to a market as the network of these cooperative relationships (Buchanan 1964, p. 218). This specific institutional framework might be contrasted with the political arena as an alternative institution for demanding and supplying goods and services (cf. Buchanan 1999, p. 3). Nevertheless, despite the concentration on the market side of exchanges, at least the distribution of property rights in the Faustmann model must be resolved by collective consensus on the constitutional stage of politics. This construction guarantees that the owner of the land is the relevant subject of the analysis. In this way, it should be emphasized that the classical assumptions underlying the Faustmann model (cf. Section 3.3.1) imply specific ranges of application since the rules among individuals are fixed just as the physical characteristics (Buchanan and Brennan 2000, p. 21). Any hypothesis of this analysis may thus only be applied to situations within these spheres of human interaction.

Among the questions which arise within the thus defined limits of the problem area are: Which incentives motivate forest owners to thin forest stands? Under which conditions are forest stands left untreated? When and how are thinnings conducted? Which trees will be removed under what condition? How often are thinnings conducted and how much timber can be expected to be removed at each harvest? How will forest owners adjust the regeneration and the rotation age when thinnings become more or less profitable? How will changes in the prices affect the optimal cutting regime? How do forest owner respond to unanticipated changes during the rotation period? Hopefully, the present study contributes to the elucidation of some of these important questions.

Earlier proposed solutions to the thinning problem of profitable timber production often focus on numerical solutions. Among these, whole-stand timber growth models (e.g. Kilkki and Väisänen 1969; Brodie et al. 1978), stage-structured timber growth models (e.g. Haight 1987; Solberg and Haight 1991) and individual tree growth models are employed (cf. Hyytiäinen and Tahvonnen 2002, p. 274). Depending on growth specifications, the latter may be separated into distance-independent (e.g. Roise 1986; Cao et al. 2006) and distance-dependent (e.g. Pukkala et al. 1998; Pukkala and Miina 1998; Vettenranta and Miina 1999) growth models (cf. Hyytiäinen et al. 2005, p. 120). While the numerical approaches share the advantage of yielding concrete solutions, their applicability is restricted to the empirical values employed. In many of these studies, thinnings increase the intertemporal income from timber production. This raises the suspicion that thinnings are defined to be profitable by the underlying assumptions.

Analytical solutions have been advanced by means of dynamic programming (e.g. Schreuder 1971) and optimal control formulations of the thinning problem (e.g. Näslund 1969; Clark and Munro 1975; Clark and De Pree 1979; Cawrse et al. 1984; Betters et al. 1991; Clark 2005, p. 39 ff.). These studies offer solutions to the simultaneous determination of optimal thinnings and the optimal rotation age. It can be shown that optimal thinnings follow a continuous, singular path. As a result, thinnings are intensified until the increase of the revenues from thinnings are equal to the decrease of the timber value at the final harvest.

The present study attempts to follow a different approach by concentrating on thinnings as a result of differing optimal harvest ages of the trees in an even-aged forest stand. Hence, the focus lies on a direct extension of the Faustmann model (cf. Chapter 3) to include thinnings. In this way, the assumption of the independence of the harvest ages in the Faustmann model is relaxed and the implications analyzed. This approach and its deductions are new contributions to forestry economic science since they offer concrete propositions of whether and how thinnings are expected to be conducted in forest stands. The main advantage over other approaches lies in the consistent extension of the Faustmann model as the theoretical basis of the economic approach towards the explanation of human interactions concering forests in a market environment. The first approach into this direction is certainly due to Johann Heinrich von Thünen, who constructed a forest thinning theory (cf. Thünen 1875, 2009) within the equilibrium setting of the isolated state (cf. Thünen 1842, 1966).

The present study is organized as follows: first, a simple theory of timber growth in forest stands is developed which allows an application to the abstraction level of the investment models employed (Chapter 2). The extension of the Faustmann model to include thinnings and two further specifications are developed in Chapter 3. This chapter also contains the specifications of the physical and rule-based assumptions underlying the solution. In the analysis in Chapter 4, those logical propositions are deduced from the investment model which are necessary for the answers to the posed questions above. These problems comprise the elaboration of the relevant range of thinnings, the optimal thinning regime and the comparative static analysis of the model. The propositions are then discussed against the background of the problem and its demarcation and conclusions are drawn (Chapter 5). Finally, the study summarized in Chapter 6.

2 Timber Growth Theory

Thinnings offer the opportunity to gain access to the control of density in a forest stand through the harvest of trees. Typically, timber growth is assumed to be density-dependent, i.e., the change in the timber volume is dependent on the current stock size (cf. Conrad and Clark 1987, p. 62). The term "density", though, is not used consistently in forestry science (Zeide 2005). Depending on the problem, density might refer to the stem number, the basal area, the timber volume, the biomass or various density indices. However, in order to derive relevant propositions concerning the influence of thinnings on the profitability of timber production, it is necessary to evaluate the impact of the removal of some trees on the remaining trees. In order to meet the abstraction level of the following investment models (cf. Chapter 3), a simple growth model is required which offers guidance within the complex structures of forest stands and allows to deduce concrete hypotheses about the basic relationships of timber production. Due to the lack of such a qualitative model, this chapter tries to outline density-dependent timber growth in even-aged and pure forest stands with the help of some basic theories of natural science.

In view of the problem to define density, the proposed model follows a different approach. It analyzes the relationship between the stand age and the initial density, which is the number of plants at the beginning of a rotation cycle. In this way, density – however defined (Zeide 2005) – is the consequence of the magnitudes of both these characteristics. In effect, density is introduced into timber growth by the initial density. The initial density is convenient for several reasons. First, it is a tangible, easily assessable and logical variable as it represents a concrete action of forest management. Second, all commonly applied density measurements arrange the initial densities at the moment of the establishment in the same order as long as the regenerated trees are all equal, i.e., those stands with the highest number of trees at the beginning of the rotation are inevitably the densest stands at that moment, which makes it meaningful to speak of dense stands. Third, the initial density is directly related to thinnings as both focus on the relevant actions: to purposefully increase or to decrease the number of trees in the stand.

Timber growth, or the change in the timber volume over the age, denotes a growth function as it describes the temporal behavior of the timber volume in the system of a forest stand. Density-dependent timber growth might then comprise both positive and negative relationships between simultaneously growing trees. The proposed model, however, focusses solely on the negative relationship which might be termed competition as this term is frequently defined by its harmful effects on the involved individuals (cf. Begon et al. 1990, p. 197). The restriction of the diverse mutual interdependencies between trees solely to the negative influences on the timber volume, though, might only be acceptable in pure stands as trees of the same species might not occupy different ecological niches thus competing for the same niche. In mixed stand, mutually reinforcing timber growth might evolve when different tree species are intermingled in specific ways.

A basic assumption might be necessary in order to justify timber growth at all. If it is assumed that trees maximize their individual fitness by producing many, healthy and widespread seeds within their species-specific reproductive strategy which are capable of establishing maximal reproductive offspring in turn, timber growth might be derived thereof. In order to produce and disseminate these seeds, trees must develop large and high crowns with a large photosynthetic mass in comparison to their neighboring trees which guarantees to utilize enough resources for a large number of healthy seeds that can be easily spread by wind or animals from their raised position on the tree. The productive mass, again, can only be supplied and supported by timber. It should be noted that this assumption might not hold for mixed forest stands (Pretzsch 2009, p. 340 f.).

2.1 The Homogenous Stand

Homogeneous timber growth or a homogeneous forest stand is defined as equal timber volumes of all trees in the stand at equal ages. As a consequence, all trees necessarily share equal timber increments and equal impacts upon each other as, otherwise, the timber volumes will diverge at some age. Trees might be growing homogeneously when they are all of the same age, of equal genetic constitution, are growing on equally fertile land and in an equal climate, are evenly distributed and treated.

Timber production focusses on the timber volume of a stand Q as the means to generate an income stream. Since a forest stand is composed of different trees, the timber volume of a stand is the sum of the volumes of each tree q^i belonging to the stand, i.e.,

$$Q = \sum_{i=1}^{n} q^i,$$ [2-1]

where n is the number of trees (stem number) in the stand, and i is the index for each tree. In a homogeneous stand, every tree has exactly the same timber volume. The calculation of the timber volume thus simplifies to

$$Q = nq^i.$$ [2-2]

Accordingly, the timber volume of stand depends on the stem number and the volume of one tree.

2.1.1 Stem Number

The stem number is the number of trees in a stand. It is the direct consequence of the initial density. The latter is the number of trees when a stand is established. While the alternatives to establish a stand are numerous – e.g. planting, sowing, natural regeneration, or even coppicing – it is irrelevant for the purpose in this paragraph how the trees have been established. In the homogeneous stand, only the number of plants matters since all trees are distributed evenly and are of the same age, which might not apply necessarily to all regeneration techniques.

Just after a stand has been established, the stem number is equal to the initial density. Thereafter, though, both might diverge. Due to natural regeneration from seeds of adjacent stands or even from trees of the same stand, the stem number might increase. Since these phenomena are highly dependent on the vicinity of a stand, they have to be excluded from this analysis in order to focus on the basic relationships occurring in all stands. On the other hand, the stem number might also decrease after initial stand establishment. The reasons for the death of trees are even more various; e.g., fungal or insect infestation, browsing damage, forest management operations, vandalism, air pollution, etc. Since all these sources are again highly dependent on the vicinity of a stand, they have to be excluded. The only relevant source for the displacement of trees that is solely influenced by the processes within a stand is density-dependent mortality. This phenomenon, which is also referred to as natural mortality or self-thinning (Zeide 1985), might reduce the stem number during the development of the stand (cf. Weiskittel et al. 2011, p. 139 ff.). For convenience, density-dependent mortality will be simply termed mortality hereafter.

As long as mortality has not occurred, the stem number equals the initial density. After mortality has taken place, the stem number is less than the initial density. The relation between stem number and density-dependent mortality in forests has been examined first by Reineke (1933). In his empirical study, he found indications of an exponentially decreasing relationship between the number of trees per unit area and their average diameter in fully-stocked and untreated stands. As a result, he derived an age-independent density index thereof. Although his investigations gave rise to controversial discussions, the index is widely accepted in forest mensuration science (Zeide 2005). Various studies have since then observed similar relationships for different tree species, locations and site qualities (e.g. Zeide 1995; Pretzsch 2000; Inoue et al. 2004).

Analogous observations were made by Yoda et al. (1963). They found decreasing weights of herbaceous plants with an increasing number of plants. Consequently, they derived the "-3/2th power law of self-thinning" which describes the relation between the average living plant mass and the plants per unit area. The analyses of both Reineke (1933) and Yoda et al. (1963) can be transformed into each other by substituting plant mass for diameter (White 1981); in this way, the former is special case of the latter.

Eventually, the reciprocal changes in tree size and tree number have been derived allometrically (cf. Mohler et al. 1978; Tang et al. 1994; Zeide 2004; Pretzsch and Biber 2005). As the size of the trees in a fully-stocked stand increases, the number of plants has to decrease in order to supply enough space for each individual. By the same token, it is possible to show a very similar relationship between equal volumes of balls and their numbers when lying close to each other on a fixed area, on the one hand, and the number and size of trees, on the other (Pretzsch 2009, p. 406 ff.). The size of a tree may be either measured in volume or living plant mass as these units are isometrically related (White 1981).

If it is necessary, then, for trees to grow in size over the age, the number of trees has to decrease once the stand is fully stocked, i.e., once a maximum of photosynthetically active mass is reached. The positive change in the size of a tree with respect to the age is assumed to be given for all relevant situations. Necessarily, trees increase their diameters as long as they are alive when old sapwood cells are no longer functional. Their capabilities to achieve this, though, are limited. Single trees and untreated stands exhibit a typical growth pattern (Assmann 1970): beginning to grow on a low level with an increasing acceleration, culminating at some age, and growing with a decreasing acceleration afterwards. At some point, they might even reduce their size due to decay. However, this stage of the development of a stand can be ignored in an analysis of profitable timber production.

In a perfectly homogenous stand, however, there is no reason why single trees should be eliminated due to density-dependent mortality. Since the timber growth of all trees is equal in every aspect, no tree has any advantage

over any other tree which could then be expanded (proportional or disproportional) to depress their neighbor until his death. Therefore, all trees
would be dying off simultaneously when the small photosynthetic mass is
unable to maintain the growing metabolism of the tree. For instance, this
stagnation phase in timber growth can be observed in comparatively dense
and almost homogeneously growing stands (e.g. pine plantations) where the
timber increments tend to cease. In consequence, density-dependent mortality has to be assumed as exogenously given and to take place on a regularly
distributed random basis at any age in which some crucial density factor (defined as a combination of tree size and number of trees) is reached. Mortality
thus occurs when a stand enters a zone of imminent competition-mortality
(Drew and Flewelling 1977). According to the self-thinning line (White
1981), mortality takes place in the initially densest stands first while it never
occurs in comparatively sparse stands (Valentine 1988).

In summary, the stem number in this approach is determined by the age s of
the stand and its initial density m, i.e.,

$$n = n(s, m). \qquad\qquad [2\text{-}3]$$

2.1.2   Diameter

The second determinant of the timber volume of a homogeneous stand is the
timber volume of a single tree q^i. Since the growth of a tree is a phenomenon
in a highly complex ecosystem, myriads of factors influence the formation of
timber (cf. Oliver and Larson 1996, p. 21 ff.). In order to focus the analysis on
the relevant aspects, the factors which are working independently of the vicinity of a stand have to be isolated.

In forest mensuration science the timber volume of a tree is typically calcu-
lated as (cf. West 2009, p. 36)

$$q^i = \hat{h}^i * (d^i)^2 * \pi/4 * b^i, \qquad\qquad [2\text{-}4]$$

where $\hat{h}^i$ is the height of the tree, d^i its stem diameter and b^i a taper reduction factor which reduces the volume of a cylinder generated by $\hat{h}^i$ and d^i to the volume equivalent of the shape of the tree. According to this calculation, the diameter, the height, and the taper reduction factor are three basic determinants of the tree volume. This separation is convenient as the last two determinants are only negligibly affected by the initial density.

As various empirical observations (e.g. Altherr 1966; Petersen and Spellmann 1993; Mäkinen and Hein 2006) as well as analytical considerations (e.g. Oliver and Larson 1996, p. 335; Pretzsch 2009) reveal, tree height is hardly affected by the initial density. Although there is a tendency of higher trees in more densely planted stands, the differences are insubstantial and negligible for the purposes of this work. Nevertheless, in actual forest stands, this proposition does only apply to the dominant height, which is the average height of the tallest trees in a stand. In homogeneous stands, though, all trees are of the same height. The shape of a tree, on the other hand, is definitely influenced by the initial density (Assmann 1970, p. 57ff). If comparing the shape of solitary grown trees with those cultivated in densely planted stands, the differences are obvious, especially for deciduous trees. However, the reduction factors might be assumed to be of equal magnitude as the timber increments are only differently distributed on the stem such that solitarily growing trees grow thicker stem bases but thinner upper sections. In this way, the independence of the taper reduction factor from the initial density must follow from the limited supply of resources.

In consequence, since tree height and reduction factors are constant over varying initial densities, they will not affect the qualitative differences between stands of the same age. Only the diameter is influenced by both age s and initial density m of a stand, i.e.,

$$d^i = d^i(s, m). \tag{2-5}$$

Since this analysis only focuses on growing trees, the diameter increases with an increasing age of a stand. Furthermore, it is irrelevant where the diameter is evaluated as long as it is the same for all trees and the taper reduction factor is adjusted.

The relationship between stand density and tree diameter might be explored with the help of the pipe model theory. Although elaborated analytically by Shinozaki et al. (1964a; 1964b), empirical observations of the relationship between diameter increment and assimilation organs are already made by Pressler (1865) or Huber (1928). Pressler (1865) pointed out that the area of timber increment in any part of the stem is proportional to the leaf area above this part. Shinozaki et al. (1964a) found a linear relation between the weight of assimilation organs and the weight of non-photosynthetic tissue. They concluded that every unit of assimilation mass is supported by a unit of non-photosynthetic tissue. Therefore, the form of a plant is a consequence of the assemblage of unit pipes.

In the case of trees, Shinozaki et al. (1964a) made another important observation. For any part of a tree which lies below all photosynthetic organs, the constant ratio between photosynthetic and non-photosynthetic mass does not hold. They understood this phenomenon as the consequence of the accumulation of disused pipes. Pipes can become inoperable for various reasons; e.g., senescence, embolisms, seasonal activity. Since a tree is unable to dispose disused pipes, they are preserved in the interior of the stem as heartwood, and, therefore, do not share a functional relationship with the active photosynthetic organs. Consequently, the relationship is only valid for the active non-photosynthetic tissue of the tree, the sapwood. Numerous empirical observations (e.g. Kaufmann and Troendle 1981; Withehead et al. 1984; Kimmins 1987; Vertessy et al. 1995; Scott et al. 1998; Eckmüller and Sterba 2000) underlie this conclusion for various tree species and sites. It also severed as a basement for several, more sophisticated growth models (e.g. Waring et al. 1982; Valentine 1985; Deleuze and Houllier 1995; Mäkelä 2002).

As a tree grows in size, newly formed sapwood cells have to support newly formed photosynthetic organs as well as they have to replace disused pipes. This "secondary growth" (Nultsch 2001, p. 296) forms the timber volume as a byproduct of the need to grow. It might be interrupted periodically or irregularly due to climatic changes. The important implication for this study, however, is that the stem diameter, as a fundamental determinant of the timber volume, is a function of the photosynthetic mass of the tree. Since this mass is concentrated on the crown surface, the potential of a tree to produce timber volume is dependent on the potential to expand its crown.

The space which the crown of a tree can possibly occupy is called the potential growing area (Assmann 1970, p. 101), or area potentially available (Brown 1965). Naturally, this concept might be extended to include the rhizosphere. However, as both areas are correlated, it is not necessary for the purposes of this work. Since the trees in a homogeneous stand are of equal height, this might also be reduced to the potential ground coverage of a tree. In the clear setting of a homogeneous stand, each tree has the same potential ground coverage which determines its diameter growth because the trees are distributed evenly over the area. Therefore, higher initial densities offer limited opportunities for crowns to expand freely. As the trees grow in size, each tree occupies more and more of its potential growing area until it is filled out and the canopy is thus closed.

The potential growing area, though, is not a homogenous source of resources. This is particularly evident when comparing a nearly solitarily growing tree and one out of a closed stand. A further extension of the potential growing area will hardly promote the timber increment of the former, but will substantially raise the increment of the latter. Furthermore, the effect varies with the tree height. The detailed reaction of trees due to enlargements of the potential ground coverage cannot be evaluated here; it is not even necessary. The only relevant aspect is that the function relating enlarged growing areas to increment reactions is strictly monotonically increasing within the competitive domain; i.e., a comparatively larger growing area gives rise to both higher crown surfaces and timber volumes if solitary growth is ruled out. Without any competition between the trees of the stand (solitary growth), the increment reaction function has to be constant because genetics and site constraints make it impossible to convert more resources into photosynthetic mass. The threshold of competition was empirically analyzed, for instance, for Loblolly pine (*Pinus taeda* L.) by Clason (1994) or Zhang et al. (1996), and for Norway spruce (*Picea abies* L.) by Lässig (1991).

Eventually, lower initial densities give rise to thicker tree, and *vice versa*, if trees compete for resources. By inference, solitary growth maximizes the timber volume of a single tree as it is provided with the maximal possible supply of resources that a site can offer and the tree can utilize. The same analysis remains valid for other diameters, such as branch diameter. Therefore, all branch diameters respond equally to the stem diameter to changes in the growing area as they represent, along with their associated photosynthetic mass, small trees within the tree.

2.1.3   Stand Volume

As equation [2-2] reveals, the timber volume of a stand depends on the stem number n and the timber volume of a tree of the stand q^i. In turn, both determinants vary with the age and the initial density. Therefore, it holds that

$$Q = Q(s, m). \tag{2-6}$$

According to the argumentation in the preceding sections, higher stem numbers and larger diameters increase, lower stem numbers and smaller diameters reduce stand volume, *ceteris paribus*. However, higher initial densities give rise to higher stem numbers but smaller diameters since they reduce the potential growing area, and *vice versa*. Therefore, both effects on the stand volume are related oppositely with respect to the initial density (cf. von Thünen 1875, p. 90; 2009, p. 92). In order to evaluate which effect prevails when, the conclusions of the preceding paragraphs have to be put together.

At the beginning of a rotation period, the stem number effect predominates inevitably. The timber volume advantage of a denser stand is then proportional to the stem number advantage. A stand with twice as many stems has twice the timber volume since all trees have the same stem volume. This proportionality will hold as long as the trees grow solitarily. Especially for comparatively low initial densities, this will persist for quite some time. Eventually, as the trees grow in height and width, they will begin to influence each other through the competition for resources. In initially dense stands, this will happen sooner after the regeneration. For initially less dense stands, the boundary of solitary growth is reached disproportionately sooner after dense stands since the trees grow in height thus influencing each other through sidelight. Forest stands at sites of low soil quality or growing in a disadvantageous climate, though, might inevitably grow solitarily as some scarce resources restrict the availability of those which are more abundant. When competition begins, trees cannot maintain their solitary growth as the availability of resources is limited. As worked out in the preceding section, smaller potential growing areas will result in less rapidly growing diameters. Therefore, the stem number advantage diminishes over the age.

Nevertheless, as long as no tree is displaced, the stem number effect will dominate; i.e., higher initial densities lead to higher stand volumes. In order to produce more timber volume, any tree in a less dense stand has not only to outgrow a tree of a denser stand, but, moreover, has to compensate for the increments of the additional trees in the denser stand. For a stand with twice as many trees, each tree has to compensate for the growth of two trees in the less dense stand. Since one tree has the same potential growing area of two trees in this case, the denser stand can benefit from the availability of the resources in a shorter time since initially denser stands reach the age of canopy closure earlier. According to the pipe model theory, more timber is produced by two trees on the same area than by one tree since the area is filled out in a shorter time by photosynthetic mass, which demands the formation of additional sapwood cells. This holds as long as the trees are distributed evenly and the additional timber volume due to the enlarged occupation of the potential growing area is not increasing at an increasing rate, which can be dismissed due to the increasingly negative influence of competition. As a consequence, more trees will yield a larger timber volume other things being equal.

Given that denser stands are capable of utilizing the available resources more comprehensively in the same time, they reach the crucial density factor for mortality earlier. Comparing two stands with adjacent initial densities before mortality has taken place, the denser stand comprises a higher stem number, smaller diameters and a larger timber volume due to the prevailing stem number effect. When a tree in the denser stand is displaced due to density-dependent mortality, both stands have the same stem number. Since the trees of the initially less dense stand have larger diameters due to the larger potential growing area, they henceforth combine a larger timber volume. While mortality subsequently reduces the stem number in both stands alternately, the initially less dense stand will maintain the comparative advantage although the difference might be very small. In summary, after mortality has taken place, the diameter effect predominates. In this way, density-dependent mortality marks the point of "crossover" (Oliver and Larson, 1996, p. 339) between the timber volumes of initially more and less dense stands. It should be emphasized that the preceding argument refers to the standing timber volume or stumpage volume. The reasoning is different for the total growth performance. Since the latter comprises the standing timber volume as well as the timber volume of dead (and removed) trees, the reasoning implies a higher total growth performance for initially denser stands as the potentially available resources of the stand can be utilized more comprehensively.

Since the emphasis lies on the timber volume, it can be inferred that stands with a higher timber volume realize a higher mean increment over the period. Timber increments of trees and untreated forest stands, however, follow a distinct course (Pretzsch 2009, p. 395 f.). Originating from negligible amounts, the increments culminate at an age depending on the ecological characteristics of the tree species as well as site characteristics, and decrease afterwards. The location of the inflection point of the timber volume, though, is also dependent on the initial density. For a solitarily growing tree, the age of increment culmination is determined by its genetics and the site characteristics. At the same time, this is the highest possible age for the culmination of a stand. Once a tree is in competition for resources, it cannot maintain the increasing growth of a solitarily growing tree before the culmination. If it is assumed that the competition in an even-aged stand intensifies monotonically from its beginning on, the inflection point of the growth function moves to younger ages the denser a stand is initially established.

The way competition intensifies with increasing age decides when the increment culminates. Because competition begins weakly with the struggle for sidelight, the solitary increments are at first not influenced significantly. Accordingly, the inflection point is only shifted slightly towards younger ages. The denser a stand is initially established, the earlier competition begins, and the earlier the increment will thus be reduced substantially. This, again, has to lead to an even earlier increment culmination. While the competition intensity depends on the age and the initial density, the increment culmination follows a very similar pattern as the threshold of mortality, and has to be located somewhere in its surrounding.

Out of this, it cannot be concluded when the timber volume increment culminates exactly, particularly, because height increment, basal area increment, timber volume increment, and others, all seem to peak at different ages (Pretzsch 2009). But it is sufficient to assume for this study that the timber volume increment reaches its maximum before the threshold of mortality. This is surely the case for stands with a relatively low initial density because density-dependent mortality might never occur. On the other hand, mortality and increment culmination for initially dense stands might nearly coincide. Therefore, once mortality has taken place, timber volume increment is decreasing over the age.

2.1.4 Thinnings

Thinnings are defined as the removal of trees without subsequent regeneration. They thus necessarily influence the timber volume of the stand. Principally, thinnings extend the potential growing area of the remaining trees in the same way as a reduction of the initial density. In contrast to the latter, though, the effect of a thinning is less pronounced than the effect of a reduction in the initial density for three reasons. First, for equal harvest ages, the time period for the response is shorter. Second, older trees might be less responsive due to senescence and growing metabolisms (Nyland 2002, p. 389 ff.). In general, younger trees respond to changes more intensely than older ones. And third, the growing areas of the remaining trees might be irregularly shaped due to the fixed positions of the trees. These growing areas are inhomogeneous sources of resources since their availability decreases with rising distance from the stem (cf. Section 2.1.2). However, if it is assumed that the reaction of the remaining trees due to the enlargement of the potential growing area in even-aged stands is equable, a basic reaction pattern can be established.

By definition, thinnings reduce the stem number of a stand. Therefore, they reduce the stand volume proportionally to the number of trees removed at the age of the thinning. Since the potential growing area is enlarged, the diameter growth of the remaining trees is forced hereafter due to crown expansions, which will increase the stand volume in turn. In the set up homogenous stand, any thinning extends the potential growing area of all remaining trees equally, i.e., in the same way as reductions of the initial density. This surely conflicts with observable evidences since the established trees are bound to their location. However, if thinnings are assumed to be conducted on a regularly distributed basis, this assumption might serve as an acceptable heuristic. As a consequence, any thinning shifts the thresholds of competition and mortality exposed in the preceding paragraphs evenly towards older and initially denser stands (Vanclay 1994, p. 182 ff.). The intensity of these shifts is dependent on the intensity of the thinning, defined here as the number of trees removed. Equally, more frequently conducted thinnings will gradually shift the thresholds. The more and the more often trees will be removed and the stronger the reaction of the remaining trees, the more the threshold will be shifted. Therefore, the intensity and the frequency of thinnings have the same qualitative effect since both enlarge the potential growing area when they are increased.

Independent of the stand age, the frequency, and the intensity, any thinning thus shifts the thresholds of competition and mortality (Powers et al., 2010). The thresholds may be delayed or the stand is put back before the threshold when it has already passed them by. If younger stands react more intensely to enlargements of the potential growing space, the thresholds will be shifted only slightly in comparison to older stands since the latter are incapable of utilizing the enlarged resource availability. The more often and/ or the more intense thinnings are conducted, the more the thresholds are shifted because the potential growing spaces are more enlarged.

In consequence, thinnings might reduce or increase the stand volume at the rotation age in the same way as the initial density. The total growth performance, on the other hand, cannot be increased with the aid of thinnings for the same reasons and within the same range as for the initial density (cf. Section 2.1.3). The opportunity to increase the stand volume is only given for combinations of rotation ages and initial densities that lay beyond the threshold of mortality. If stands in which mortality will never take place are thinned, the stand volume will be reduced. Conversely, if thinnings are conducted in stands where mortality will potentially occur, stand volume can be increased, in particular, if the thinning regime is conducted in a way that the threshold of mortality shifted by the thinnings intersects the rotation age. If, for instance, thinnings are conducted as anticipating mortality, the thinned stand will have a higher stand volume for the same reason as a lower initial density. However, even a more intense as well as a less intense thinning regime might increase the stand volume. Within the range of solitary growth, any thinning will have no effect on the development of the timber volume of the remaining stand first since trees are not influencing each other.

In summary, thinnings can be interpreted as *ex post* reductions of the initial density. While their influence is thus qualitatively equivalent, the intensity of the responses of the remaining trees is weaker since older trees might be less responsive and the time interval of possible adaptions on the part of the remaining trees might be shorter.

2.1.5   Summary

The preceding argumentation is summarized in Figure 2.1. In order to distinguish the relations shown here from functional relations, Figure 2.1 is set up in the form of a table where the rows and columns are continuously merged into each other. In this way, the columns on the horizontal axis denote increasing stand ages while the rows on the vertical axis denote increasing initial densities. The functional relationship behind Figure 2.1 could be expressed by a third axis which denotes the timber volume. Since the growth processes were evaluated only qualitatively, this axis remains indefinite. Therefore, Figure 2.1 is the two-dimensional projection of the timber volume as a function of the stand age and the initial density.

In Figure 2.1 the dotted curve delimits the range of solitary growth. Within this range, trees do not influence each other's growth; i.e., no competition takes place. The higher the initial density, the earlier competition will arise. For some very low initial densities, trees will not influence each other at any

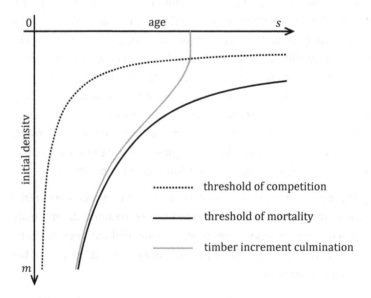

Figure 2.1 A graphical representation of the timber growth theory

age. Within the solitary range, the differences in the timber volume at equal ages are proportional to the differences in the initial densities. The solid black curve shows the threshold of density-dependent mortality. It is not the self-thinning line since it does not illustrate the development of the stem number. As worked out in the preceding paragraphs, the higher the initial density, the earlier mortality will occur. The threshold of mortality marks the boundary of the ranges of predominance of the stem number, and of the diameter effect respectively. Beyond the threshold of competition and before mortality takes place, stands originating from a higher initial density accumulate a higher timber volume though disproportionately less than the differences in the stem numbers since each tree comprises a smaller timber volume. Finally, the solid grey line marks the timber volume increment culmination. As argued above, it is independent of the initial density when trees grow solitarily. When competition takes place, it peaks the earlier the initially denser the stands are established.

Thinnings will cause both the competition and the mortality threshold to shift away from the origin. The extent of the shift depends on the age of the trees and the number of trees removed. The younger the trees and the more trees are removed, the greater the shift. In order to maximize the timber volume at a particular age, the initial density must be chosen in a way that mortality has just not occurred yet. The reverse, though, is not true. The age that maximizes timber volume for a particular initial density may emerge before or after mortality has taken place depending on the genetically and site specifically determined age of decay.

Figure 2.1 might equally be converted into a set of derivatives describing the changes in the timber volume of a single tree and the stand due to changes in the age and the initial density. Table 2-1 shows the corresponding changes within the ranges of solitary and competitive growth and of competitive growth with mortality separated by the thresholds of competition and of mortality. Most changes in the variables, especially those within the mortality range, are unambiguous. The second order derivatives with respect to the initial density are negative when solitary growth maximizes the timber growth of single trees. Ambiguous, though, are the changes in the timber in-

Table 2-1 Changes in stem diameter, tree volume and stand volume due to changes in the stand age and the initial density as derived from the timber growth theory

change	solitary	threshold of competition competitive	threshold of mortality mortality
$\partial d^i/\partial s$	> 0	> 0	> 0
$\partial^2 d^i/\partial s^2$	$\lessgtr 0$	$\lessgtr 0$	< 0
$\partial d^i/\partial m$	$= 0$	< 0	< 0
$\partial^2 d^i/\partial m^2$	$= 0$	< 0	< 0
$\partial^2 d^i/\partial s\partial m$	$= 0$	$\lessgtr 0$	≥ 0
$\partial q^i/\partial s$	> 0	> 0	> 0
$\partial^2 q^i/\partial s^2$	$\lessgtr 0$	$\lessgtr 0$	< 0
$\partial q^i/\partial m$	$= 0$	< 0	< 0
$\partial^2 q^i/\partial m^2$	$= 0$	< 0	< 0
$\partial^2 q^i/\partial s\partial m$	$= 0$	$\lessgtr 0$	≥ 0
$\partial Q/\partial s$	> 0	> 0	> 0
$\partial^2 Q/\partial s^2$	$\lessgtr 0$	$\lessgtr 0$	< 0
$\partial Q/\partial m$	$= Q/m$	> 0	< 0
$\partial^2 Q/\partial m^2$	$= 0$	< 0	< 0
$\partial^2 Q/\partial s\partial m$	$= 0$	$\lessgtr 0$	≥ 0

crements within the ranges of solitary and competitive growth without mortality. As for the increment culmination with respect to the stand age (cf. Section 2.1.3), the cross derivatives of age and initial density might be positive, negative or zero as Chang (1983, p. 269) has pointed out. After density-dependent mortality has initiated, the cross-derivative must be non-negative as, otherwise, timber volumes of stands with different initial densities might cross again.

2.2   The Heterogeneous Stand

In contrast to homogeneous forest stands, trees in heterogeneous forest stands might accumulate different timber volumes at equal ages. However, heterogeneous stands must be restricted to even-aged and pure stands in this analysis. While the basic ecological theories employed above do equally apply the mixed and uneven-aged stands, additional characteristics would have to be considered. For instance, mutually reinforcing timber growth might occur in mixed stands when different tree species occupy different ecological niches, whereas density-dependent mortality might occur at any age in the uneven-aged stand.

Naturally, differing timber volumes are rooted in differing timber increments. In the simplest case, a heterogeneous stand consists of two parts or classes of trees with differing timber volumes but equal tree timber volumes within each class. At the other end of the range, each tree of the stand might grow to a different timber volume over an equal period. In general, the timber growth of trees in a stand differs for every individual. Especially in stands with long rotation periods growing on untreated sites, the differences might become significant.

The cause of heterogeneity of timber volumes might be traced back to dissimilar genetic constitutions and irregular site characteristics (Assmann 1970, p. 41). Differences in both features lead to unequal timber growth. By inference, in order to establish homogeneous stands, clones with identical genetics are usually planted on ploughed, leveled and fertilized land. Differences in the genetic constitution might refer to both the variability within a species as well as to the different genetic codes of different species. The site characteristics comprise all effects on the macro- and the micro-level of the soil, the climate and the biotic factors. Before competition between the trees arises, these factors determine the differences between the trees. Since the concept of the forest stand implies more or less uniform growth conditions, the differences between the trees in pure stands are usually small during the phase of solitary growth.

In the course of the stand development, however, competition between the trees of the stand might reinforce the differences. Trees with a competitive advantage, which must not be correlated with the timber volume, become predominant since the advantage enables them to expand their access to the contested resources at the expense of the inferior growing tree. Depending on the competing species, the intra-competitive advantage might be proportionally or disproportionally expandable due to size-symmetries (cf. Schwinning and Weiner 1998). As a result, the differences between the trees increase after the threshold of competition has been exceeded until some trees die off due to the absence of available resources. Typically, the higher the initial density, the less shade-tolerant the species and the more vigorous the tree growth, the faster stands are differentiating and the earlier mortality takes place (Oliver and Larson 1996, p. 217 ff.).

In this way, density-dependent mortality is an endogenous process due to the heterogeneous growth conditions and not an exogenously determined detail as in the homogeneous stand. Since trees respond to competition with reductions of the diameter increments first while reductions of the height increments do not follow until competition has intensified severely (cf. Section 2.1.2), density-dependent mortality is anticipated by tree height differentiations. Hence, dominant tree height of the remaining trees remains unaffected by competition. However, for heterogeneous stands, the development of the timber volume might not be deduced in an equally clear way. For instance, density-dependent mortality might occur earlier in initially less dense stands when some trees are more handicapped as in the initially denser stand. In this case, the self-thinning line is a curve rather than a straight line in the log-scale (Weller 1987). This transmission of the age-independent Reineke- and Yoda-model to an age-dependent stand development model, though, might cause misunderstandings as White (1981) has pointed out. He clearly differentiates between the self-thinning line of Reineke (1933) and the competition-density effect; the latter leads to curves that approach the self-thinning line asymptotically. In the homogeneous stand both curves coincide. Nevertheless, the basic development remains valid if extreme forms of competitive disadvantages are excluded. In these cases, the area might have to be separated into two stands with more uniform growth conditions.

In heterogeneous stand, though, different thinning methods might be applied (cf. Nyland 2002, p. 407 ff.). These different methods of selecting trees to be thinned surely influence the development of the timber volume and the thresholds of competition and mortality (cf. Bradford and Palik 2009; Powers et al. 2010). Removals of more vigorously growing trees and the deliberate protection of less vigorously growing trees might shift the mortality and competition thresholds less intense than the consequent removal of inferior trees. Since a comparatively large amount of resources is released when trees with large crowns are removed, the timber volume of a stand is reduced in the short and the long run as not all of these resources might be utilized by the remaining trees due to large gaps in the canopy. In this way, thinnings from above lead to lower stand volumes than thinnings from below.

2.3 Empirical Reference

From the proposed timber growth model, hypothesis might be derived which can be tested against the background of empirical observations. These might include:

1. Density-dependent mortality marks the crossover effect (Oliver and Larson 1996, p. 339). As a consequence, higher initial densities accumulate larger timber volumes at comparatively low stand ages while lower

initial densities lead to higher standing timber volumes in higher ages if the stands are growing untreated. In order to produce the maximal volume of standing timber, a stand has to be regenerated with an initial density where mortality has just not yet occurred when the stand is harvested.

2. The maximal total growth performance is expected in stands with the highest initial densities.

3. Thinnings increase the standing timber volume when they shift the threshold of mortality towards the harvest age. Thinnings decrease the standing timber volume when they shift the threshold of mortality away from the harvest age.

The analysis of the impact of the initial density upon the timber volume of a stand has been a subject of forestry science since its very beginning. Various empirical studies have been conducted subsequently. A serious problem of experimental plots in forestry, though, is their long time horizon. Especially in temperate climates, potential rotation ages are long in comparison to the development of the environment and the society. In order to analyze some different initial densities and their influence in combination with the rotation age under a plot design with three repetitions, two to three centuries might pass by. Within this period, not only the biological conditions can vary substantially but also the objectives of the scientists entrusted with these plots.

For instance, the social development can be witnessed by the designs of the plots. Whereas older plots of Norway spruce (*Picea abies* L.) in the nineteenth and the beginning of the twentieth century have been more concerned with comparatively high initial densities (> 3000 plants/ha; cf. Busse and Jaehn 1925), more recent experiments analyze less dense stands (- 2000 plants/ha; cf. Mäkinen and Hein 2006).

Another serious methodical problem for the testing of deductive hypotheses is the wide range of different thinning regimes that have been carried out in the course of the evolution of the various experimental stands. Since virtually no thinning regime has been applied identically in two otherwise similar plots, the obtained results can hardly be compared. The results are therefore often interpreted in the light of a somehow defined density index (e.g. in terms of basal area, stem number, etc.). As a consequence of the large number and variety of empirical studies, some results seem to contradict each other.

Due to the different empirical observations, two patterns of timber growth have been put forward (cf. Oliver and Larson 1996, p. 339). According to the constant-yield effect, different initial densities eventually converge to the same timber volumes. The crossover effect, on the other hand, supposes that timber volumes of stands developed out of lower initial densities eventually exceed the timber volume of initially denser stands. Similarly, various growth models which have been parameterized with compilations of empirical data (simulators) develop crossover patterns (O'Hara and Oliver 1988). The different growth patterns have been summarized by Zeide (2001), who seemed to found evidence for both optimal and increasing relationships between density – however defined (Zeide 2005) – and timber volume.

The proposed timber growth theory provides an explanation for the crossover effect (cf. Oliver and Larson 1996, p. 339) while it conflicts with the constant-yield effect. Accordingly, the stem number effect dominates the diameter effect as long as all trees remain in the stand such that higher initial densities lead to higher timber volumes. The occurrence of density-dependent mortality, on the other hand, marks the crossover. Hence, optimal and increasing relationships between density and timber volume are not mutually contradictory, but are dependent on the initial density and the thinning regime in relation to the stand age. When comparing stands evolving out of a range of relatively high initial densities at relatively advanced ages, i.e., after mortality has taken place, initially less dense stands accumulate a larger stand volume. On the contrary, when evaluating the same stands at comparatively younger ages, or a range of stands of comparatively low initial densities, the maximum timber volume is found in the initial denser stands. For stands in a range of comparatively low initial densities, the maximal stand volume will be located in the denser and unthinned stands for a wide range of stand ages. In this way, the model offers an explanation for both increasing timber volumes with an increasing density (e.g. for low initial densities, short rotation ages and/ or severely thinned stands) as well as an optimum course (e.g. for high initial densities, long rotation ages and moderately thinned or unthinned stands).

For Norway spruce (*Picea abies* L.), for instance, these interrelations might be reconstructed. In the nineteenth and at the beginning of the twentieth century, experimental plots were typically designed with relatively high planting densities. When these plots were evaluated at a relatively young age, the initially densest stands comprised the highest timber volume (Vanselow 1950). Some years later, the maximal stand volume had been found in less dense stands while mortality has taken place (Vanselow 1956). Similar results can be seen in the report of Busse and Jaehn (1925) and for European beech (*Fagus sylvatica* L.) in Schwappach (1911). In more recent experiments, relatively low plantings densities have been investigated. In this case, the maximal volume had been found in the densest stands (cf. Kramer and Spellmann 1980; Petersen and Spellmann 1993; Mäkinen and Hein 2006). Gizachew et al. (2012) conclude that the opportunity to produce more timber with initially denser stands might be restricted to the early stand development. Nevertheless, in all observations, the highest total growth performance has been located in the densest stands. Oliver and Larson (1996, p. 339 ff.) give many more examples for the crossover effect which can be tested for the occurrence of density-dependent mortality.

Concerning thinnings, the growth model does not exclude an increasing or optimum pattern between the thinning regime and the stand volume. When initially dense stands are thinned severely or moderately at a younger or even an older age, the stand volume at a comparatively high rotation age will be higher than without thinnings. For lower initial densities, only moderate thinnings in a younger age can increase the stand volume at the same age. For short rotation ages and/ or initially less dense stands, it is not possible to increase the stand volume by thinnings. These deductions do not conflict with observations in experimental plots. For instance, as Mäkinen and Isomäki (2004) observed for comparatively low ages in Norway spruce stands, thinnings of different intensities did not increase total growth performance and equally reduced the standing timber volume slightly while increasing the mean stem diameter and reducing, i.e. postponing, natural mortality. For comparatively high stand ages of Norway spruce and European beech, by contrast, Pretzsch (2005) found both increasing merchantable timber volumes for a moderate thinning intensity and decreasing merchantable timber volumes for heavy thinning intensities compared to light thinnings. In the former case, the threshold of mortality might have been shifted towards the age of observation while, in the latter case, the comparatively intense thinning caused the threshold of mortality to be shifted away from the age of observation.

Finally, the constant-yield growth patterns (cf. Oliver and Larson 1996, p. 339) might be explained by the comparatively high degree of heterogeneity in the development of old forest stands. When the differences between the timber volumes of stands with unequal initial densities are only small, stands might seem to converge to the same timber volumes over a wide range of initial densities if extremely low densities are excluded. Since more random events might occur over longer stand ages, the differences might become indistinct in this range. In the model approach, the observation that the differences are small over wide ranges of initial densities is caused by the alternate mortality after the threshold of mortality has been crossed. Nevertheless, the growth model theory in this chapter is not restricted to these ranges but applies to any range of initial densities and stand ages.

3 Investment Model

The correct formulation of the problem of a forest owner willing to maximize his intertemporal income with the production of timber within a partial equilibrium was first presented by Martin Faustmann (1849) although his intention then has been to construct a formula for the calculation of the forest land value. The corresponding Faustmann formula combines the relevant aspects of profitable timber production by the cyclic regeneration and harvesting of an even-aged stand. In its original version, it constitutes the net present value of bare forest land, often referred to as the land expectation value (Amacher et al. 2009, p. 20) after Faustmann (LEV^F), of an infinite, periodic sequence of payments at different stand ages less the net present value of operation costs, i.e., in a modern notation,

$$LEV^F = \frac{pQ^T + \sum_{i=1}^{j} pQ^i w^{(T-t_i)} - Cw^T}{w^T - 1} - \frac{\hat{a}}{v}. \qquad [3\text{-}1]$$

Any period covers the costs for the establishment or the regeneration of the stand C at the beginning of a rotation period, the sum of the j revenues from selling the timber volume of each thinning Q^i at the stand age t_i for a net timber price p, and the revenues from the sale of the harvested stand volume Q^T at the rotation age T at the net timber price p. In this setting, all payments are prolonged to the rotation age with the help of the discount factor $w = 1 + v$, where v is the annual interest rate. If it is assumed that the sequence of payments is equal in each rotation period, the infinite chain of rotation periods is converted into a present value via a capitalization sequence. Moreover, the present value of annual costs $\hat{a}$ at the interest rate v have to be subtracted.

In deductive analyses of the profitability of timber production, the original version of the Faustmann formula [3-1] is often simplified for the derivation of a correct solution in a partial equilibrium (e.g. Samuelson 1976; Johansson and Löfgren 1985, p. 73 ff.; Amacher et al. 2009, p. 11 ff.). In order to account for the characteristics of timber growth, the discrete discounting is usually

transformed into its continuous counterpart with r as the continuous interest rate. Furthermore, annual costs are ignored since they are assumed to be
independent of the rotation age and, therefore, irrelevant for any qualitative
analysis. Finally, thinnings are ignored or assumed to be exogenously determined and constant. Therefore, the analytical land expectation value (LEV^a),
which constitutes the Faustmann model, can be written as

$$LEV^s = \frac{pQe^{-rT} - C}{1 - e^{-rT}}. \qquad [3\text{-}2]$$

In this simplified version, timber growth Q is assumed to be dependent solely
on the rotation age of the stand, i.e.,

$$Q = Q(T). \qquad [3\text{-}3]$$

This assumption serves as a heuristic approach for a general analysis. Since
timber growth is the consequence of myriads of factors in a highly complex
ecosystem (Oliver and Larson 1996, p. 41 ff.), all negligible influences are excluded in order to focus on the relevant aspects of profitable timber production (cf. Johansson and Löfgren 1985, p. 77f.). The fact that timber volume
production is dependent on the corresponding area (cf. Chapter 2), i.e. $Q = Q(T, A)$ where A is the area, is simply regarded for by setting $A = 1$
(Johansson and Löfgren 1985, p. 75). In this way, the Faustmann model pro-
vides an analysis on the forest stand level as the basic management unit
within forests which is assumed to offer more or less uniform management
conditions.

3.1   Thinning model

The LEV gives the net present value of bare forest land per unit area. Accordingly, the timber growth function Q gives the timber volume per unit area.
As described in Paragraph 2.1, this stand volume is composed of the trees
which are growing on the unit area. Depending on the extent of the unit area,

the timber volume may thus comprise one or many trees. If the timber volume of a single tree is denoted by q^i and the number of trees is n, the stand volume in [3-3] might be rewritten as

$$Q = \sum_{i=1}^{n} q^i. \qquad\qquad\qquad [3\text{-}4]$$

All trees in the Faustmann model [3-2] share the same rotation age due to the uniform discounting of the timber value. As a consequence, all tree volumes vary with the rotation age which is inevitably equal for all trees. This assumption might be dropped when an individual rotation age t_i is attached to each tree. Since the growth of each tree might depend on the harvest of previously removed trees (cf. Section 2.1.4), the growth function is given by

$$q^i = q^i(t_1, \dots, t_i), \qquad\qquad\qquad [3\text{-}5]$$

where the subscripts to the harvest ages indicate the temporal order of harvests such that t_i is the harvest age of the corresponding tree i denoted in the superscript of the growth function q, and $t_1, \dots, t_{i-1}$ are the harvest ages of potentially previously harvested trees with $t_i \geq t_{i-1}$.

If the harvest of a previously cut tree is followed by the regeneration of the bare patch of land it leaves behind, the subsequent stand is uneven-aged. By contrast, thinnings are commonly understood as the harvest of trees without the opportunity or the objective to regenerate the created gap. In order to analyze the profitability of the latter, the sum in [3-4] has to be inserted in the Faustmann model [3-2] while considering [3-5] and the necessary adjustment of the discount factors, i.e.,

$$LEV = (1 - e^{-rt_n})^{-1}\left(\sum_{i=1}^{n} pq^i e^{-rt_i} - C\right), \qquad\qquad [3\text{-}6]$$

where n is the number of trees within a rotation period. In this way, the harvest of trees prior to the rotation age is possible while the stand is regenerated only after the harvest of the last tree at the rotation age t_n. This construction necessarily results in an even-aged stand if the regeneration is ensued with equally old trees.

The objective of any forest owner willing to maximize his intertemporal profits (cf. Chapter 1) is then

$$\max_{t_1,\dots,t_n} LEV(t_1,\dots,t_i) \quad s.t. \quad t_i \geq t_{i-1} \quad and \quad t_1,\dots,t_n \geq 0. \qquad [3\text{-}7]$$

Restricting all growth functions to be at least twice continuously differentiable, [3-7] constitutes a nonlinear optimization problem which can be solved by employing the Lagrange function L in connection with the Kuhn-Tucker necessary maximum conditions. These take the general form of

$$\max_{t_1,\dots,t_n,\lambda_1,\dots,\lambda_n} L = (1 - e^{-rt_n})^{-1}\left(\sum_{i=1}^{n} pq^i e^{-rt_i} - C\right)$$
$$+ \sum_{i=1}^{n} \lambda_i\left(t_i - t_{t_{i-1}}\right) \qquad\qquad [3\text{-}8]$$

$$\frac{\partial L}{\partial \lambda_i} = t_i - t_{t_{i-1}} \geq 0 \quad \lambda_i \geq 0 \quad and \quad \lambda_i \frac{\partial L}{\partial \lambda_i} = 0 \qquad [3\text{-}9]$$

$$\frac{\partial L}{\partial t_i} = \frac{\partial LEV}{\partial t_i} + \lambda_i \leq 0 \quad t_i \geq 0 \quad and \quad t_i \frac{\partial L}{\partial t_i} = 0. \qquad [3\text{-}10]$$

If $t_i > t_{i-1}$, then $\lambda_i = 0$ and thus $\partial L / \partial t_i = \partial LEV / \partial t_i$. If $t_i = t_{i-1}$, the problem is either transformed to classes of trees with equal harvest ages and the same form of maximum conditions as [3-10] or it is reduces to the Faustmann model [3-2] as all trees share the same rotation age. Hence, as long as the constraint is satisfied, the attention is restricted to interior solutions which can be analyzed with unconstrained maximization techniques.

For an unconstrained maximum, the necessary condition is given by

$$\frac{\partial LEV}{\partial t_1} = \cdots = \frac{\partial LEV}{\partial t_k} = \cdots = \frac{\partial LEV}{\partial t_n} = 0\Bigg|_{(t_1^*,\ldots,t_n^*)}, \qquad [3\text{-}11]$$

where $t_1^*, \ldots, t_n^*$ are the optimal harvest ages of each tree which maximize the LEV. The necessary condition for a maximum thus comprises n equations with n unknowns.

In order to obtain the required partial derivatives, the objective function [3-6] can be rewritten as

$$LEV = (1 - e^{-rt_n})^{-1}\left(pq^1 e^{-rt_1} + \sum_{i=2}^{k-1} pq^i e^{-rt_i} + pq^k e^{-rt_k} \right.$$

$$\left. + \sum_{j=k+1}^{n-1} pq^j e^{-rt_j} + pq^n e^{-rt_n} - C \right). \qquad [3\text{-}12]$$

Accordingly, the relevant components of the first order partial derivatives are separated. The latter take the general form of

$$\frac{\partial LEV}{\partial t_1} = (1 - e^{-rt_n})^{-1}\left(pq_{t_1}^1 e^{-rt_1} - rpq^1 e^{-rt_1} + \sum_{j=2}^{n} pq_{t_1}^j e^{-rt_j} \right) \quad [3\text{-}13]$$

$$\vdots \qquad\qquad \vdots \qquad\qquad \vdots$$

$$\frac{\partial LEV}{\partial t_k} = (1 - e^{-rt_n})^{-1}\left(pq_{t_k}^k e^{-rt_k} - rpq^k e^{-rt_k} \right.$$

$$\left. + \sum_{j=k+1}^{n} pq_{t_k}^j e^{-rt_j} \right) \qquad [3\text{-}14]$$

$$\vdots \qquad\qquad \vdots \qquad\qquad \vdots$$

$$\frac{\partial LEV}{\partial t_n} = (1 - e^{-rt_n})^{-1}\left(pq_{t_n}^n e^{-rt_n} - rpq^n e^{-rt_n} - rLEV e^{-rt_n} \right), \quad [3\text{-}15]$$

where the subscripts to the functions indicate partial derivatives and k might be any tree between 1 and n.

In order to ensure that a point which satisfies condition [3-11] is a maximum, the second order condition for a local maximum has to be fulfilled. It requires that

$$|H_1| < 0; |H_2| > 0; |H_3| < 0; \dots; (-1)^n |H_n| > 0|_{(t_1^*, \dots, t_n^*)}, \qquad [3\text{-}16]$$

where

$$|H_i| \equiv \begin{vmatrix} \dfrac{\partial^2 LEV}{\partial t_1^{\,2}} & \cdots & \dfrac{\partial^2 LEV}{\partial t_1 \partial t_i} \\ \vdots & \ddots & \vdots \\ \dfrac{\partial^2 LEV}{\partial t_i \partial t_1} & \cdots & \dfrac{\partial^2 LEV}{\partial t_i^{\,2}} \end{vmatrix} \qquad [3\text{-}17]$$

is the ith leading principal minor of the Hessian determinant to [3-11] and n is the number of trees in the stand. According to Young's theorem, the matrices in the sufficient condition [3-16] are symmetric.

The corresponding second order partial derivatives are

$$\frac{\partial^2 LEV}{\partial t_1^{\,2}} = (1 - e^{-rt_n})^{-1}\left(-2rpq_{t_1}^1 e^{-rt_1} + r^2 pq^1 e^{-rt_1} \right.$$
$$\left. + \sum_{j=1}^{n} pq_{t_1 t_1}^j e^{-rt_j} \right) \qquad [3\text{-}18]$$

$$\vdots \qquad\qquad \vdots \qquad\qquad \vdots$$

$$\frac{\partial^2 LEV}{\partial t_k^{\,2}} = (1 - e^{-rt_n})^{-1}\left(-2rpq_{t_k}^k e^{-rt_k} + r^2 pq^k e^{-rt_k} \right.$$
$$\left. + \sum_{j=1}^{n} pq_{t_k t_k}^j e^{-rt_j} \right) \qquad [3\text{-}19]$$

$$\vdots \qquad\qquad \vdots \qquad\qquad \vdots$$

$$
\begin{aligned}
\frac{\partial^2 LEV}{\partial t_n{}^2} = (1 - e^{-rt_n})^{-2}\big[&(pq_{t_n t_n}^n e^{-rt_n} - 2rpq_{t_n}^n e^{-rt_n} \\
&+ r^2 pq^n e^{-rt_n} - rLEV_{t_n} e^{-rt_n} \\
&- r^2 LEV e^{-rt_n})(1 - e^{-rt_n}) \\
&- (pq_{t_n}^n e^{-rt_n} - rpq^n e^{-rt_n} \\
&- rLEV e^{-rt_n})re^{-rt_n}\big].
\end{aligned}
\qquad [3\text{-}20]
$$

The partial cross-derivatives take the general form of

$$
\frac{\partial^2 LEV}{\partial t_k \partial t_1} = (1 - e^{-rt_n})^{-1}\left(-rpq_{t_1}^k e^{-rt_k} + \sum_{j=k}^{n} pq_{t_k t_1}^j e^{-rt_j}\right)
\qquad [3\text{-}21]
$$

$$
\vdots \qquad\qquad \vdots \qquad\qquad \vdots
$$

$$
\frac{\partial^2 LEV}{\partial t_n \partial t_k} = (1 - e^{-rt_n})^{-1}\big(pq_{t_n t_k}^n e^{-rt_n} - rpq_{t_k}^n e^{-rt_n} \\
- rLEV_{t_k} e^{-rt_n}\big).
\qquad [3\text{-}22]
$$

The presented model is restricted in its domain and in the range of its functions. The corresponding constraints are addressed in Section 3.3.2.

3.2 Model extensions

The model presented in the preceding Paragraph 3.1 is subject to almost all of the analyses in this study in order to guarantee a consistent approach to the problem. It allows concentrating on the basic influence thinnings exert on the land expectation value. However, while the model is reduced to work in this simplified manner, many extensions might be devised. Extensions, though, only refer to modifications of the model structure while specifications of parameters are made in the analysis directly.

The model is extended into two directions. On the one hand, the initial density is introduced as an endogenous variable. Since the initial density directly controls the number of trees in a stand, this extension might be vital for a comprehensive analysis of the influence of thinnings on the profitability of timber production. On the other hand, the model is extended towards a more general uneven-aged setting. In view of the various opportunities to specify and restructure the model, these two extensions should be understood as examples. The extensions shall explore both the limits and the internal logic of the model.

3.2.1   Initial Density

In the original Faustmann formula [3-1] as well as in its analytical version [3-2], the initial density is implicitly regarded as a constant in the regeneration costs. However, the production theory of the previous chapter indicates the dependency of the timber volume on the initial density (cf. Section 2.1.1). For instance, initially denser stands offer numerous opportunities to thin as more trees are available. A general model including the scale of the investment was proposed by Hirshleifer (1970, p. 91). Its explicitly forestry related counterpart has been presented by Hyde (1980, p. 52). Lastly, Chang (1983) comprehensively analyzed the influence of the initial density on the profitability of timber production with a more specific model. In his work, he extended the Faustmann model [3-2] for the planting (as an initial) density m as an endogenous variable.

In this sense, the timber volume of each tree is varying with the initial density m next to its own and all previously conducted harvest ages within the rotation period, i.e.,

$$q^i = q^i(t_1, \dots, t_i, m).$$ [3-23]

Furthermore, the initial density helps determining the regeneration costs since (cf. Chang 1983, p. 268)

$$C = mC_v + C_f,$$ [3-24]

where C_v are the variable regeneration costs while C_f represent the fixed regeneration costs. Applied to the thinning model [3-6], this results in

$$LEV^m = (1 - e^{-rt_n})^{-1} \left(\sum_{i=1}^{n} pq^i e^{-rt_i} - mC_v - C_f \right). \qquad [3\text{-}25]$$

Yet, the initial density exerts another influence in this model as it determines the numbers of trees available for harvest. Put differently,

$$n = f(m). \qquad [3\text{-}26]$$

In order to eliminate n in the upper bound of the summation index, [3-25] is rewritten as

$$LEV^m = (1 - e^{-rt_n})^{-1} \left(\sum_{i=1}^{k} pq^i e^{-rt_i} + \sum_{j=k+1}^{n} pq^j e^{-rt_j} \right.$$
$$\left. - mC_v - C_f \right). \qquad [3\text{-}27]$$

The second term in the second bracket might constitute the class of trees cut at the rotation age. In this case, $t_j = t_{j+1} = \cdots = t_n$. If these trees grow identically, then

$$\sum_{j=k+1}^{n} pq^j e^{-rt_j} = [n - (k+1) + 1] pq^n e^{-rt_n}$$
$$= (n - k)pq^n e^{-rt_n} \qquad [3\text{-}28]$$

with the result that

$$LEV^m = (1 - e^{-rt_n})^{-1} \left(\sum_{i=1}^{k} pq^i e^{-rt_i} + (n - k)pq^n e^{-rt_n} \right.$$
$$\left. - mC_v - C_f \right). \qquad [3\text{-}29]$$

With the initial density, the equation system constituting the necessary condition [3-11] is extended by one equation, i.e.,

$$\frac{\partial LEV^m}{\partial t_1} = \cdots = \frac{\partial LEV^m}{\partial t_n} = \frac{\partial LEV^m}{\partial m} = 0 \Big|_{(t_1^*,\ldots,t_n^*,m^*)}, \qquad [3\text{-}30]$$

where m^* is the optimal initial density in the sense that it constitutes a maximum of the LEV. The corresponding first order conditions are analogous to [3-13] - [3-15] with the difference that each timber growth function refers to [3-23] and the LEV to [3-29]. The additional first order condition is given by

$$\frac{\partial LEV^m}{\partial m} = (1 - e^{-rt_n})^{-1}\left(\sum_{i=1}^{k} pq_m^i e^{-rt_i} + n_m pq^n e^{-rt_n} \right.$$
$$\left. + (n - k)pq_m^n e^{-rt_n} - C_v \right). \qquad [3\text{-}31]$$

The third term in the second bracket might now be added to the summation index as

$$\frac{\partial LEV^m}{\partial m} = (1 - e^{-rt_n})^{-1}\left(\sum_{i=1}^{n} pq_m^i e^{-rt_i} + n_m pq^n e^{-rt_n} - C_v \right) \qquad [3\text{-}32]$$

such that the trees which have been previously assumed to belong to the rotation class do not necessarily share the same harvest ages and growth functions to any further extent.

The second order sufficient condition requires that

$$|H_1| < 0; |H_2| > 0; |H_3| < 0; \ldots; (-1)^{n+1}|H_{n+1}| > 0 \big|_{(t_1^*,\ldots,t_n^*,m^*)} \qquad [3\text{-}33]$$

with

$$|H_{n+1}| \equiv \begin{vmatrix} \dfrac{\partial^2 LEV^m}{\partial t_1{}^2} & \cdots & \dfrac{\partial^2 LEV^m}{\partial t_1 \partial t_n} & \dfrac{\partial^2 LEV^m}{\partial t_1 \partial m} \\ \vdots & \ddots & \vdots & \vdots \\ \dfrac{\partial^2 LEV^m}{\partial t_n \partial t_1} & \cdots & \dfrac{\partial^2 LEV^m}{\partial t_n{}^2} & \dfrac{\partial^2 LEV^m}{\partial t_n \partial m} \\ \dfrac{\partial^2 LEV^m}{\partial m \partial t_1} & \cdots & \dfrac{\partial^2 LEV^m}{\partial m \partial t_n} & \dfrac{\partial^2 LEV^m}{\partial m^2} \end{vmatrix} \qquad \text{[3-34]}$$

as the symmetrical Hessian determinant to equation system [3-30] and n as the number of trees in the stand.

In the same way as for the necessary condition [3-30], the corresponding second order partial derivatives [3-18] - [3-20] have to be adjusted for [3-23] and [3-29] whereas the additional derivative is

$$\frac{\partial^2 LEV^m}{\partial m^2} = (1 - e^{-rt_n})^{-1} \left(\sum_{i=1}^{n} pq^i_{mm} e^{-rt_i} + n_{mm} pq^n e^{-rt_n} \right). \qquad \text{[3-35]}$$

The additional cross partial derivatives take the general form of

$$\frac{\partial^2 LEV}{\partial t_1 \partial m} = (1 - e^{-rt_n})^{-1} \left(-rpq^1_m e^{-rt_1} + \sum_{j=1}^{n} pq^j_{t_1 m} e^{-rt_j} \right) \qquad \text{[3-36]}$$

$$\vdots \qquad\qquad \vdots \qquad\qquad \vdots$$

$$\frac{\partial^2 LEV}{\partial t_n \partial m} = (1 - e^{-rt_n})^{-1} \left(pq^n_{t_n m} e^{-rt_n} - rpq^n_m e^{-rt_n} \right.$$
$$\left. - rLEV^m_m e^{-rt_n} \right). \qquad \text{[3-37]}$$

Constraints to the range of the functions are specified in Section 3.3.2.

3.2.2 Uneven-aged stands

The thinning model [3-6] is constructed in order to exclude the opportunity to regeneration the patch of bare land which a thinning leaves behind instantaneously. This is probably the common understanding of thinnings. If, on the other hand, the harvest of less than all trees of a stand is followed by the regeneration of bare patches, the resulting stand will be uneven-aged. This opportunity can be included in the model by integrating the capitalization sequence into the sum, adjusting the indices and assigning the prorated regeneration costs to each tree, i.e.,

$$LEV^u = \sum_{i=1}^{n} (1 - e^{-rt_i})^{-1} (pq^i e^{-rt_i} - C_i), \qquad\qquad [3\text{-}38]$$

where LEV^u is the land expectation value of an uneven-aged stand. Here, n is the number of trees which might possibly occur in different rotation periods; however not all of these might be present at the same time. The index i denotes the temporal order of the harvests.

In this setting, the harvest of a tree might not only influence the remaining trees but also those trees which are not even present in the stand at the harvest since the potential harvest might shape their growing conditions differently. Therefore,

$$q^i = q^i(t_1, \dots, t_n). \qquad\qquad [3\text{-}39]$$

The necessary condition is analogous to [3-13] - [3-15]. The first order partial derivative with respect to the harvest of any tree k takes the general form of

$$\frac{\partial LEV^u}{\partial t_k} = (1 - e^{-rt_k})^{-1} \big[pq_{t_k}^k e^{-rt_k} - rpq^k e^{-rt_k}$$

$$- r(1 - e^{-rt_k})^{-1}(pq^k e^{-rt_k} - C_k)e^{-rt_k}\big]$$

$$+ \sum_{i \in (i \in n | i \neq k)} (1 - e^{-rt_i})^{-1}\big(pq_{t_k}^i e^{-rt_i}\big). \qquad\qquad [3\text{-}40]$$

The derivation of the second order condition has been omitted as the focus in this study lies on the even-aged thinning model [3-6].

3.3 Assumptions and Constraints

The maximization of the objective function, namely [3-7], is only meaningful in a clearly defined setting. Otherwise, any derived hypothesis is empty in substance as auxiliary hypotheses might in principle immunize any refutation. With the intention of preventing any inconsistencies within the interpretation of the results, assumptions are necessary which delimit the scope of the analysis. Additionally, constraints on the domain and range of the involved functions should guarantee the reference to observable problems.

3.3.1 The Faustmann Laboratory

In order to provide a correct solution in a precisely delimited environment, the analysis of this study is conducted in a partial equilibrium of forestry. In contrast to a general equilibrium of all markets, a partial equilibrium is restricted to a specific range of human interactions; i.e., in this study, interactions concerning profitable timber production. The analysis follows then a *ceteris paribus* approach (cf. Marshall 1922, p. 363; Samuelson 1983, p. 19); namely, all other things are held constant. The precise separation between the artificially created equilibrium and the remainder of human interactions is ensured by exogenously determined prices of selected goods. The prices evolve and the goods are discovered by factors outside of the partial equilibrium. Whether the prices and goods are determined within an equilibrium or not is of no interest as long as the selected interactions satisfy the prescribed assumptions. Surely, the severity of the assumptions might demand equally severe assumption for other interactions. The partial equilibrium implies, moreover, that changes in the selected interactions do not affect all other interactions. Naturally, both implications restrict the applicability of the results obtained if they are not considered against the background of the underlying assumptions.

An equilibrium is understood here as a state of the model where the selected variables show no inherent tendency to change (cf. Machlup 1958, p. 9). The selection of the variables defines the range of application. Propositions can only be raised with regard to these variables. The inherence guarantees the balance of internal forces and, by inference, the invariability of external forces. Owing to the lack of change, the analysis of the states of equilibria is often referred to as statics.

The equilibrium per se, though, serves only as a preliminary. With its artificially created invariability, it ensures, on the one hand, that without exogenously introduced changes in the parameters the state of balance remains, and, on the other hand, that with exogenously introduced changes these are the sole cause of the adjustment process followed by the unbalancing interference. The assurance that all adjustment processes have been considered is provided by the attainment of a new equilibrium with no further adjustments due to the missing inherent tendency to change. The created equilibria thus serve as "methodological devices" (Machlup 1958, p. 5) to guarantee that other factors are absent and all changes are considered. As a consequence, the empirically testable and interpersonally traceable hypotheses (Popper 2002b, p. 23) in a world of unanticipated changes refer to the change within this mentally constructed laboratory. In this way, the equilibrium theory infers that movements towards equilibria are the consequence of exogenous disturbances. Since disturbances occur within a time period, the hypotheses might equally, in the sense of physics, be termed dynamic (Machlup 1959; Popper 2010, p. 424).

The "Faustmann laboratory" (Deegen et al. 2011, p. 363) is the suitable partial equilibrium for the analysis of timber production. It is built with the help of the classical stringent, or "heroic" (Samuelson 1976, p. 470), assumptions (cf. Johansson and Löfgren 1985, p. 74f.; Amacher et al. 2009, p. 18), which are capable of producing a correct solution. First, capital markets are assumed to be operating in a way that the forest owner is able to transform any income stream at an exogenously given, known and certain market interest rate (perfect capital market). This construction allows separating the consumptive preferences of the forest owner from his productive opportunities to generate income (cf. Fisher 1930, p. 125 ff.). In this way, the analysis can be concentrated solely on the productive optima with no further regards to consumption as the realization of the overall optimum becomes an independent two-stage process (Hirshleifer 1970, p. 63).

Second, and analogously to first, the market of forest land is assumed to operate in a way that the forest owner can buy and sell any amount of forest land (usually termed perfect land market). This assumption ensures that the forest owner has an incentive to regenerate a forest stand even if the rotation period exceeds his personal life span, for he has always the opportunity to sell the forest stand at the price that equals the capitalized income stream generated from the management of the forest stand. In combination with the perfect capital market, which together guarantee the partial equilibrium analysis, the time horizon of the regeneration investment becomes infinite as the highest forest price can only be obtained if all future rotation periods are considered. In this precise sense, the Faustmann model might be termed sustainable as the continual production of timber for all future generations is regarded by the current management. Besides, since the basic object of the analysis is the forest stand as a more or less uniform management unit, external factors, such as interdependencies with other stands of the forest or with surrounding land in general, have to be assumed not to affect the optimality through constructions as the linear forest (Johansson and Löfgren, 1985 p. 112 ff.) or the normal forest viewed in the long-term stationary state (Tahvonen and Viitala 2006).

Third, prices for timber and regeneration inputs as well as timber yields and interest rates are known and their rate of change is constant. In this study, though, the rate of change is assumed to be zero such that prices and yields are constant. The Faustmann laboratory, though, has not been abandoned when prices and yields change at a constant nonzero rate over the age since the relative investment alternatives remain equal in every period. While these aspects surely affect the results of the analysis, they do not violate the partial equilibrium setting. Biotechnological improvements or timber price increases at constant rates might thus be included with no loss of generality. Constantly changing prices and the infinite time horizon lead to equally long rotation periods which allow capitalizing all future income streams by an infinite sequence where any addition of more rotation periods become obsolete. It should be emphasized, however, that these classical assumptions imply specific ranges of application since the rules among individuals are fixed just as the physical characteristics (cf. Chapter 1).

Due to the prescribed knowledge of future investment situations, the thus set up model is purely deterministic. In empirical studies, however, it is observed that the investment parameters change over time (cf. Duerr 1993, p. 346). While stochastic relationships might introduce risk as a model component, uncertainty remains. In this way, age and time are synchronized in the Faustmann model such that the influence of unanticipated changes is held constant. The stringent assumptions thus convert the complex social problem into a computable maximization algorithm. From Chapter 1, though, it must be recalled that this clear result of the action theory serves only as a preliminary for the actually relevant interaction theory.

In the environment of the partial equilibrium of profitable timber production, the objective of the rational forest owner constructed in Chapter 1 becomes the maximization of the land expectation value through the variation of the harvest ages as the selected, relevant variables. Since the land expectation value is the present value of the income stream generated by timber production, higher land expectation values offer a larger set of consumption opportunities for the forest owner (cf. Chapter 1).

3.3.2   Range Constraints

The presented model [3-6] is restricted by plausibility constraints concerning its domain and the range of the involved functions and investment parameters. As already comprised in the model derivation, cf. [3-7], the domain of the land expectation value is restricted to positive harvest ages. While it is virtually impossible for t_i to take negative values as this implies to harvest trees in the past, a harvest age of zero would amount to the constantly planting and immediate harvesting of the stand. If this could earn a positive rent, the wealth obtainable with the production of timber is infinite. Since infinite wealth is not observed, [3-7] can only yield an interior solution.

In the analysis, the investment parameters p, r and C are assumed to be positive. Though net timber unit revenues might be negative, p refers to the timber price. Variable harvest costs are introduced later. Similar, the continuous, real market interest rate r is assumed to be positive. Although situations might be constructed where zero or even negative real interest rates would probably be prevailing, this would imply that there are no or just disadvantageous opportunities for the alternative use of resources. These situations, however, are rare and seldom long lasting (cf. Fisher 1930, p. 89). Moreover, the LEV is assumed to be positive at least in its maximum. Otherwise, timber production is unprofitable on the whole and thus not an object of economic investigation.

The timber volume functions are assumed to be at least twice continuously differentiable. Furthermore, they are restricted to positive values and increments, i.e.,

$$q^i > 0; \quad \frac{\partial q^i}{\partial t_i} > 0. \qquad\qquad [3\text{-}41]$$

Accordingly, the analysis is limited to positive timber volumes as there are no negative quantities. Besides, the timber increment of each tree is required to be positive. Though negative growth is conceivable due to decay, only positive increments are relevant in a homogeneous stand as there is always the opportunity to regenerate the stand in order to yield positive increments or timber production as a whole is unprofitable.

Next to these plausibility constraints, the timber volume function is restricted by the sufficient condition [3-11]. Explicitly, it imposes limitations to the growth and thinning impact accelerations, which, however, are quite complex. In effect, these limitations demand that trees are not harvested before the age where the value growth acceleration outweighs the interest on the value increment adjusted for thinning impacts. In order to illustrate this point, the thinning model [3-6] might be simplified to comprise only two harvest ages (indicated by LEV^b) such that the maximization problem is given by

$$\max_{t_k, t_n} LEV^b = (1 - e^{-rt_n})^{-1} \left(\sum_{i=1}^{k} pq^i(t_k)e^{-rt_k} \right.$$

$$\left. + \sum_{j=k+1}^{n} pq^j(t_k, t_n)e^{-rt_n} - C \right) \qquad [3\text{-}42]$$

$$s.t. \quad t_n \geq t_k \quad and \quad t_k, t_n \geq 0,$$

where t_n is the rotation age and t_k is the potential thinning age. The necessary condition for a maximum comprises then a system of two equations, i.e.,

$$\frac{\partial LEV^b}{\partial t_k}\bigg|_{(t_k^*, t_n^*)} = \sum_{i=1}^{k} pq_{t_k}^i - r\sum_{i=1}^{k} pq^i + \sum_{j=k+1}^{n} pq_{t_k}^j e^{r(t_k - t_n)} = 0 \qquad [3\text{-}43]$$

$$\frac{\partial LEV^b}{\partial t_n}\bigg|_{(t_k^*, t_n^*)} = \sum_{j=k+1}^{n} pq_{t_n}^j - r\sum_{j=k+1}^{n} pq^j - rLEV = 0, \qquad [3\text{-}44]$$

with t_k^* as the optimal thinning age and t_n^* as the optimal rotation age. In order to constitute a maximum, both second order derivatives must be negative semidefinite. Employing the necessary condition [3-43] and [3-44], these demand that

$$\frac{\partial^2 LEV^b}{\partial t_k \partial t_k}\bigg|_{(t_k^*, t_n^*)} = \sum_{i=1}^{k} q_{t_k t_k}^i - r\sum_{i=1}^{k} q_{t_k}^i + \sum_{j=k+1}^{n} pq_{t_k t_k}^j e^{r(t_k - t_n)}$$

$$+ r\sum_{j=k+1}^{n} q_{t_k}^j e^{r(t_k - t_n)} \leq 0 \qquad [3\text{-}45]$$

$$\Leftrightarrow \frac{\sum_{i=1}^{k} q_{t_k t_k}^i + \sum_{j=k+1}^{n} q_{t_k t_k}^j e^{r(t_k - t_n)}}{\sum_{i=1}^{k} q_{t_k}^i - \sum_{j=k+1}^{n} q_{t_k}^j e^{r(t_k - t_n)}} \leq r$$

$$\frac{\partial^2 LEV^b}{\partial t_n^2}\bigg|_{(t_k^*, t_n^*)} = \sum_{j=k+1}^{n} q_{t_n t_n}^j - r\sum_{j=k+1}^{n} q_{t_n}^j \leq 0$$

$$\Leftrightarrow \frac{\sum_{j=k+1}^{n} q_{t_n t_n}^j}{\sum_{j=k+1}^{n} q_{t_n}^j} \leq r. \qquad [3\text{-}46]$$

Accordingly, for constant timber prices, the timber volume acceleration rate of the trees cut at the rotation age must be equal to or less than the rate of interest. For the optimal thinning ages, this acceleration rate is adjusted for the potential impact and its acceleration on the remaining trees. If these characteristics hold for every age within the relevant domain, the LEV gives rise to a strictly quasi-concave function where the point satisfying the necessary condition determines the unique global maximum. In connection with [3-41], the sufficient conditions ensure that the timber growth rates, i.e. $q_{t_i}^i / q^i$, are declining since they exceed the acceleration rates, namely $q_{t_i t_i}^i / q_{t_i}^i$.

Finally, there might be restrictions to the impact of previously conducted harvests on the timber volume of remaining trees within a rotation period. In pure stands, the impact is typically negative or at least zero, i.e.,

$$\frac{\partial q^i}{\partial t_j} \leq 0 \ \ with \ j \in \{1, \ldots, i-1\}. \tag{3-47}$$

The restriction of the diverse mutual interdependencies between trees solely to the negative influences on the timber volume, though, might only be acceptable in pure and competitive stands as trees of the same species might not occupy different ecological niches thus competing for the same niche. In this way, competition is defined by its harmful effects on the involved individuals (cf. Begon et al. 1990, p. 197). Without competition, the impact is zero. In mixed stands, on the other hand, it is conceivable that the different competition strategies of different tree species might improve each other's timber growth during the joint growth phase. Assumption [3-47], though, is not necessary for the analysis below (cf. Section 5.1.2). Nevertheless, it helps to concentrate on the relevant interdependencies in homogeneous forest stands.

4 Analysis

In a partial equilibrium, defined by the stringent assumptions concerning market performances as well as future price and growth developments (cf. Paragraph 3.3), the optimal management regime for an income maximizing forest owner is indicated by the simultaneous satisfaction of the equation system [3-11] which constitutes the first order condition for a maximum of the *LEV*. In this way, the optimal harvest ages define the optimal cutting regime. However, in order to derive tangible indicators, this chapter tries to explore the implications for relevant problems of timber production, which follow from the equilibrium system. These include: when to harvest a tree prior to or at the rotation age (cf. Paragraph 4.1) and under which investment situations this separation becomes relevant (cf. Paragraph 4.2); how intense or frequent trees are harvested prior to the optimal rotation age and in which order they are optimally cut (cf. Paragraph 4.3); which influence is based on the harvest costs, how many trees are initially regenerated and what is the optimal timber volume of the stand (cf. Paragraph 4.4); how do changes in the investment situation influence the optimal harvest ages (cf. Paragraph 4.5).

4.1 Optimal Harvest Ages

Optimal harvest ages are defined as the solutions to the first-order necessary condition [3-11] for a maximum of the *LEV* with respect to harvest ages assuming the second-order sufficient condition [3-16] to hold. In principle, two different forms of optimal harvest ages occur in the thinning model [3-6]. On the one hand, the optimal harvest age of all trees which are cut instantaneously before the regeneration of the stand is termed the optimal rotation age (Section 4.1.1). One the other hand, the optimal harvest ages of all trees harvested without instant regeneration are the optimal thinning ages (Section 4.1.2).

4.1.1   Optimal Rotation Age

The optimal rotation age is determined by the optimal harvest age t_n^* of last tree within the rotation period in a maximum of the *LEV*. According to [3-11], this requires condition [3-15] to be equal to zero, which yields after rearranging

$$pq_{t_n}^n = rpq^n + rLEV.$$ [4-1]

As a result, the optimal rotation age is attained when the marginal revenue given by the value increment of the last tree to be cut equals the marginal cost given by the interest on both the timber value of the last tree and the *LEV*, with the former known as the cost of holding the trees and the latter the cost of holding the land. In a partial equilibrium (cf. Section 3.3.1), the *LEV* defines the most profitable land use as the infinite income stream generated with the production of timber. Hence, the interest on the *LEV* is equivalent to the land rent. The incentives to clear-cut a forest stand are thus given by poor value increment of the timber or high cost of either or both the timber and the land value.

Condition [4-1] bears resemblance to the Faustmann-Pressler-Ohlin (*FPO*) theorem (Johansson and Löfgren 1985, p. 80) which states the necessary condition for the maximization of the Faustmann model [3-2] with respect to the rotation age *T*, i.e.,

$$pQ_T = rpQ + rLEV^a,$$ [4-2]

which is a function of the optimal Faustmann rotation age T^*. In either condition, timber value increment is pitted against the interest on the timber value and on the land value. The distinctions between the *FPO* theorem and the maximum condition for the optimal rotation age of the thinning model [4-1], however, are threefold. On the one hand, condition [4-1] is a function of n equations in contrast to [4-2] which is only dependent on *T*. Second, condition [4-1] refers to the timber volume of a single tree q while the *FPO* theorem is determined with the help of the timber volume of the stand Q. Third, the land rent is determined by the *LEV* associated with thinning model [3-6] or the Faustmann model [3-2], respectively.

In order to reveal the differences for the location of the optimal rotation age, [4-1] and [4-2] are rewritten in terms of rates as

$$\frac{pq_{t_n}^n}{pq^n} = r + \frac{rLEV}{pq^n} \qquad\qquad\qquad [4\text{-}3]$$

$$\frac{pQ_T}{pQ} = r + \frac{rLEV^a}{pQ}. \qquad\qquad\qquad [4\text{-}4]$$

In either way, the timber value growth rate is required to equal the rate of interest in sum with the land rent in relation to the employed timber value. The value growth rates on the left hand sides both represent the change in the timber value in relation to the timber value itself. Contrary to [4-3], the *FPO* theorem captures all trees of the stand. Its corresponding value growth rate refers to the mean of the growth rates of each tree. In the case of a homogeneous stand, $Q = nq$, and $Q_T = nq_T$, cf. [3-4]. As n can thus be cancelled out of the left hand side of [4-4], the value growth rate of one tree remains as in the case of condition [4-3]. In a heterogeneous stand, the relative form of the *FPO* theorem [4-4] balances the arithmetic mean of all individual value growth rates of each tree. By way of contrast, the number of trees in the stand is relevant for the relative land rent, i.e., the second terms on the right hand side of both [4-3] and [4-4]. In the *FPO* theorem, the land rent is set in relation to the timber value of the stand, whereas in [4-3] the land rent is dispersed over the value of one tree only.

Figure 4.1 illustrates the determination of the optimal rotation age graphically with respect to [4-3]. The solid black curve represents the value growth rate. Typically, it decreases monotonically at a decreasing rate (cf. Section 3.3.2). The relative land rent might take the course of the dashed grey curve. It is equally decreasing monotonically as the land value is a positive constant determined by the maximal *LEV*, whereas the timber value increases over the relevant range (cf. Section 3.3.2). The intersection point of the value

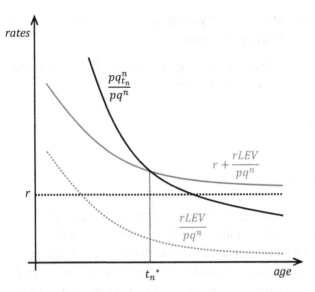

Figure 4.1 The isolated determination of the optimal rotation age

growth rate with the sum of the relative land rent and the rate of interest (solid grey curve) marks the optimal rotation age $t_n{}^*$.

4.1.1.1   Comparison of Optimal Rotation Ages

How is the optimal rotation age of the thinning model [3-6] related to the optimal rotation age with respect to the Faustmann model [3-2]? Put differently, how do thinnings influence the optimal rotation age? In order to assess the difference between both approaches, the Faustmann model [3-2] might be rewritten as

$$LEV^a = (1 - e^{-rT})^{-1} \left(\sum_{i=1}^{n} pq^i(T)e^{-rT} - C \right).$$ [4-5]

In contrast to the thinning model, all trees are harvested at the rotation age T. The *FPO* theorem [4-2] in the converted notation is then

$$\sum_{i=1}^{n} pq_T^i = r \sum_{i=1}^{n} pq^i + rLEV^a \qquad [4\text{-}6]$$

or, written in rates,

$$\frac{\sum_{i=1}^{n} pq_T^i}{\sum_{i=1}^{n} pq^i} = r + r \frac{LEV^a}{\sum_{i=1}^{n} pq^i}. \qquad [4\text{-}7]$$

If thinnings are integrated as an exogenously given, additional income stream independent of the harvest ages of the remaining trees and occurring at a given age of $t < T$, the Faustmann model in [4-5] would be extended as

$$LEV^1 = (1 - e^{-rT})^{-1} \left(\sum_{j=1}^{k} pq^j e^{-rt} + \sum_{i=1}^{n} pq^i(T) e^{-rT} - C \right). \qquad [4\text{-}8]$$

The corresponding maximum condition is adjusted to lead to

$$\sum_{i=1}^{n} pq_T^i = r \sum_{i=1}^{n} pq^i + rLEV^a + r(1 - e^{-rT})^{-1} \sum_{j=1}^{k} pq^j e^{r(T-t)}. \qquad [4\text{-}9]$$

If the additional income pq^i is positive, the optimal rotation age decreases as the cost side increases while the marginal revenues remain unchanged. Therefore, it pays to clear cut the stand more frequently in order to obtain the additional income more often. For a negative additional income stream, the optimal rotation age increases.

If thinnings occur in the form that some of the n trees are cut prior to the rotation age, i.e., at the age $t < T$, while leaving the remaining trees unaffected, for instance, through the collection of dead wood, the LEV is

$$LEV^2 = (1 - e^{-rT})^{-1} \left(\sum_{i=1}^{k} pq^i e^{-rt} + \sum_{j=k+1}^{n} pq^j(T) e^{-rT} - C \right). \qquad [4\text{-}10]$$

The maximum condition written in rates is then

$$\frac{\sum_{j=k+1}^{n} pq_{T}^{j}}{\sum_{j=k+1}^{n} pq^{j}} = r + r\frac{LEV^{2}}{\sum_{j=k+1}^{n} pq^{j}}.$$ [4-11]

In contrast to [4-7], the timber value at the rotation age comprises only $n - (k + 1)$ trees. In a homogenous stand, the value growth rate remains unaffected as the number of trees is cancelled out. In heterogeneous stands, the value growth rate in [4-11] is higher than in [4-7] if less vigorously growing trees are cut previously and lower for the thinning of more vigorously growing trees. The right hand side of [4-11] is necessarily higher if $LEV^{2} \geq LEV^{s}$ since the denominator is lower. Together, the optimal rotation age in [4-11] is less than in [4-7] for homogenous stands and equal land rents. If the land rent with thinnings decreases or the mean growth rate increases, the change in the rotation age depends on the corresponding magnitudes.

Finally, in the thinning model with interdependencies [3-6], the optimal rotation age for $n - (k + 1)$ trees is determined according to, cf. [4-3] in connection with [4-11],

$$\frac{\sum_{j=k+1}^{n} pq_{t_n}^{j}}{\sum_{j=k+1}^{n} pq^{j}} = r + r\frac{LEV}{\sum_{j=k+1}^{n} pq^{j}}.$$ [4-12]

In contrast to [4-11], [4-12] is a function of all harvest ages. However, the change in the rotation age also depends on the corresponding magnitudes. For homogeneous, competitive and equivalent stands, the value growth rate increases compared to [4-7] as they generate additional increments while they decrease the timber volume at the rotation age (for more details cf. Section 4.4.4). This tends to lengthen the optimal rotation age. However, the cost side increases likewise since the timber value is reduced while the LEV increases for profitable thinnings, which tends to shorten the optimal rotation age. In summary, the change in the optimal Faustmann rotation age due to the introduction of profitable thinnings is ambiguous and depends on the magnitude of the changes involved.

4.1.2 Optimal Thinning Age

The optimal thinning age is determined by the optimal harvest age t_k of a tree within the rotation period in a maximum of the *LEV* which is not followed immediately by the regeneration of the stand. According to [3-11], this requires condition [3-14] to equal zero, which yields after rearranging

$$pq_{t_k}^k = rpq^k - \sum_{j=k+1}^{n} pq_{t_k}^j e^{r(t_k-t_j)}. \qquad [4\text{-}13]$$

As a result, the optimal thinning age is determined by the equality of the timber value increment of the corresponding tree with the interest on the timber value of the tree as well as the present value of the impacts of a postponement of the harvest of the kth tree on all remaining trees. Since the additional impacts on the timber growth functions of all remaining trees are negative in a competitive and equivalent forest stand (cf. Section 3.3.2), the minus in front of the second term on the right hand side of [4-13] turns into a plus thus constituting the impaired growth of the remaining trees due to a postponement of the thinning into opportunity costs of not harvesting the tree. Accordingly, there might be three incentives to thin a stand: to liquidate low value increments, to earn an alternative income, to improve the growth of the remaining trees. Or *vice versa*, obtaining an additional income in terms of timber value increment is priced by the yield of the alternative investment and the impaired growth of all remaining trees.

Analogous to [4-3] and [4-4], condition [4-13] might be rewritten in rates, i.e.,

$$\frac{pq_{t_k}^k}{pq^k} = r - \sum_{j=k+1}^{n} \frac{pq_{t_k}^j}{pq^k} e^{r(t_k-t_j)}. \qquad [4\text{-}14]$$

From this perspective, the optimal thinning age is determined by the equality of the value growth rate with the rate of interest and the present value of the impacts of a postponement of the harvest on the remaining trees in relation to the value of the thinned tree, or shortly, the present value of the impact

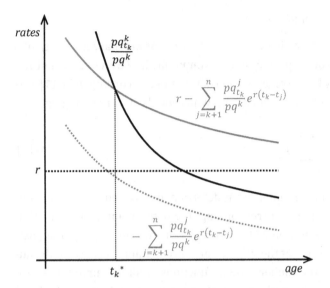

Figure 4.2 The isolated determination of the optimal thinning age

rate. A comparison of conditions [4-3], [4-4], and [4-13] emphasizes the similarity between the determinations of the optimal harvest ages. In each case, the value growth rate of a tree is balanced with the rate of interest. Additionally, the sum of all additional impacts on future harvest ages is included on the cost side.

Figure 4.2 displays the isolated determination of the optimal thinning age according to [4-14] graphically. The solid black curve represents a typical, monotonically decreasing course of the value growth rate of a tree. At some age, the growth rate intersects the rate of interest from above. The dotted grey curve shows a hypothetical course of the additional impact of a postponement of the thinning. The impact rate, thereby, decreases monotonically over the stand age for a competitive and equivalent stand as the nominator is constant or decreases while the denominator increases, cf. Section 3.3.2. In this example, the impact rate is assumed to be negative thus turning the minus in front into a plus. Added to the rate of interest, both denote the cost side of postponing the thinning of the tree illustrated by the solid grey curve.

The intersection point with the growth rate marks the optimal thinning age $t_k{}^*$. Before the intersection, the value growth rate of the tree exceeds the sum of the interest rate and the relative influence on the remaining trees. After the intersection, it pays to transform the timber value and its growing area into the next best alternative use.

4.2   The Relevance of Thinnings

Thinnings are denoted as relevant if at least one investment situation, defined by the investment parameters (cf. Chapter 1), exists for which at least one tree is cut prior to the rotation age. In contrast, thinnings are termed irrelevant when all trees share the same optimal harvest age necessarily and independent of the investment situation. In this case, the thinning model [3-6] would be reduced to the Faustmann model [3-2]. In the thinning model [3-6], though, the isolated condition for the optimal rotation age comprises only one tree, cf. e.g. [4-3]. However, the optimal rotation age in the thinning model refers to the simultaneous fulfillment of all equations in the system constituting the first-order necessary condition [3-11]. Hence, there might be more than on tree to be cut at the rotation age.

In order to analyze this implication, the first order condition for the tree to be harvested before the last tree, i.e., $n - 1$, is rewritten as, cf. [3-14],

$$\frac{pq_{t_{n-1}}^{n-1}}{pq^{n-1}} = r - \frac{pq_{t_{n-1}}^{n}}{pq^{n-1}} e^{r(t_{n-1}-t_n)}. \qquad [4\text{-}15]$$

Accordingly, the optimal second last harvest age within the rotation period is determined by the equality of the value growth rate on the left hand side and the rate of interest and the present value of the impact rate on the last tree to be cut. Isolating the rate of interest on one side of [4-3] and [4-15], both trees n and $n - 1$ are cut at the same age, i.e., $t_n{}^* = t_{n-1}{}^*$, if

$$\frac{pq_{t_n}^{n}}{pq^{n}} - \frac{pq_{t_{n-1}}^{n-1}}{pq^{n-1}} = \frac{rLEV}{pq^{n}} + \frac{pq_{t_{n-1}}^{n}}{pq^{n-1}}. \qquad [4\text{-}16]$$

Condition [4-16] exemplifies the basic incentives for conducting thinnings. In order to separate the different aspects, the implications are first analyzed with reference to the homogenous stand (Section 4.2.1) and subsequently in the more complex context of the heterogeneous stand (Section 4.2.2).

4.2.1   In the Homogenous Stand

In a homogeneous forest stand (cf. Paragraph 2.1), trees of the same age grow equally. In this case, the difference on the left hand side of [4-16] equals zero leaving [4-16] as

$$rLEV = -pq_{t_{n-1}}^n \qquad\qquad\qquad [4\text{-}17]$$

since $pq^n = pq^{n-1}$ for $t_n{}^* = t_{n-1}{}^*$ in the homogenous stand. Accordingly, the two last trees in the stand are both cut at the rotation age if the land rent equals the negative of the impact of a postponement of the thinning on the last tree.

This approach can be generalized for the homogenous stand. If the rotation class, i.e., all trees to be cut at the rotation age, comprises $n - (k + 1)$ trees, the optimal rotation age is determined by the equality of, cf. [4-3],

$$\frac{\sum_{j=k+1}^n pq_{t_j}^j}{\sum_{j=k+1}^n pq^j} = r + \frac{rLEV}{\sum_{j=k+1}^n pq^j}$$

$$\Leftrightarrow \quad \frac{pq_{t_j}^j}{pq^j} = r + \frac{rLEV}{(n-k)pq^j} \qquad\qquad [4\text{-}18]$$

while the optimal harvest age of the last tree k considered to be thinned is, cf. [4-14],

$$\frac{pq_{t_k}^k}{pq^k} = r - \sum_{j=k+1}^n \frac{pq_{t_k}^j}{pq^k} e^{r(t_k - t_j)}$$

$$\Leftrightarrow \quad \frac{pq_{t_k}^k}{pq^k} = r - \frac{(n-k)pq_{t_k}^j}{pq^k e^{r(t_n - t_k)}}. \qquad\qquad [4\text{-}19]$$

This kth tree is thus added to rotation class of the homogenous stand, i.e., $t_k^* = t_n^*$, if

$$\frac{rLEV}{(n-k)pq^j} = -\frac{(n-k)pq_{t_k}^j}{pq^k}$$

$$\Leftrightarrow \frac{rLEV}{(n-k)^2} = -pq_{t_k}^j$$

[4-20]

since $pq^k = pq^j$ for $t_k^* = t_n^*$ in the homogenous stand.

According to [4-20], the thinning and rotation classes are separated by the inversely proportional relationship between the impact of a postponement of the thinning on one of the trees in the rotation class (left hand side) and the square of the number of trees in the rotation class multiplied with the land rent (right hand side). As long as the land rent in relation to the number of trees does not fall short of the potential influence of a postponed thinning, all trees are cut at the rotation age. Conversely, all trees for which the additional thinning impact outweighs the relative land rent are thinned. Thinnings are thus relevant in the homogenous stand if the additional thinning impact exceeds the relative land rent, i.e., if

$$\frac{rLEV}{(n-k)^2} < -pq_{t_k}^j.$$

[4-21]

Accordingly, whether thinnings are conducted in a homogeneous stand or not is determined by the impact on the remaining trees, but not by the liquidation of unsatisfactory growth rates or alternative investment opportunities as these are not part of the right hand side of [4-21]. If condition [4-21] is satisfied for at least one tree in the stand, thinnings are relevant as the LEV might be increased by the harvest of trees prior to the rotation age. If the rotation class would only comprise a single tree, i.e., $k = n - 1$, condition [4-20] would be reduced to [4-17].

Condition [4-21] provides some general implications for the homogenous stand. At first, the basic prerequisites for thinnings to be relevant are interdependencies between the trees of a stand regarding their timber growth. If the harvest of a tree does not influence the timber growth of the remaining trees, i.e., if trees are growing solitarily, thinnings cannot increase the LEV in any investment situation. In this case, the right hand side of [4-21] turns zero. In order to be satisfied, either or all the interest rate, the LEV and/ or the tree number must be negative. However, since all three components are necessarily positive for any meaningful analysis of profitable timber production (cf. Section 3.3.2), [4-21] will not be satisfied for any tree. Therefore, all trees are cut at the rotation age.

Without interdependencies between the trees, the optimal thinning condition [4-14] is reduced to the solution for a single optimal rotation or duration (cf. Hirshleifer 1970, p. 82ff) as the second term on the right hand side turns zero. In this case, it pays to hold the tree until its value growth rate equals the rate of interest. However, in the homogenous stand, this implies the optimal thinning age to exceed the optimal rotation age since the latter is, according to [4-3], determined with the help of the relative land rent, which is necessarily positive (cf. Section 3.3.2). Hence, the optimal rotation age is located before the intersection point of the value growth rate and the interest rate in Figure 4.1. The difference is the same as between the optimal duration and the rotation age (Hirshleifer 1970, p. 86; Samuelson 1976, p. 481; Johansson and Löfgren 1985, p. 81). An optimal thinning age exceeding the optimal rotation age conflicts with the model approach, cf. [3-7], as regeneration would take place before the harvest of the last tree. Since in the homogenous stand growth and interdependencies between the trees are reciprocal, all trees would be cut at the rotation age.

Referring to the timber growth theory summarized in Figure 2.1, thinnings are thus irrelevant for all combinations of the initial density and the rotation age which are located within the range of solitary growth marked by the dotted curve. However, persistent solitary growth is irrelevant for profitable timber production in general. Within the range of solitary growth in a homogeneous stand, another tree planted will raise the timber volume at any age proportionally to the rise in the stem number (cf. Section 2.1.3). Hence, a doubling of the initial density would double the timber volume within this range. These correlations give rise to constant returns to scale (cf. Varian 2010, p. 341f.) since the *LEV* is then linear homogeneous with respect to the initial planting density. If q denotes the growth of a tree in dependence of the rotation age T, p the corresponding timber price, c the variable regeneration costs of one tree and m the initial density, any modification of the tree number by an arbitrary factor β will modify the *LEV* proportionately since

$$LEV^\beta(T, \beta m) = \frac{\beta mpq(T) - \beta mce^{rT}}{e^{rT} - 1} = \beta LEV^\beta(T, m). \qquad [4\text{-}22]$$

If the *LEV* is positive, any increase of the initial density will raise the *LEV* in this setting. Consequently, for any maximal *LEV*, the forest stand is regenerated at least at a density that will result in competition between the trees at the end of the rotation. With fixed regeneration costs, the same reasoning remains valid although these might lead to negative *LEV*s for low initial densities.

As a result, thinnings are invariably relevant in the homogenous stand since stands will only be managed in a way that mutual interdependencies between the trees will arise during the rotation period. However, in view of the fact that every stand development begins with a more or less long period of solitary growth, thinnings are irrelevant during these ages. When any potential postponement of a thinning has no effect on the other trees during this period, i.e., as long as the threshold of competition has not been crossed, it always pays to hold all trees in the homogenous stand as they offer growth rates considerably higher than the next best alternative. Although thinnings within the solitary growth range extend the unbounded growth for the remaining trees, the alternative of yielding a value increment during this time is superior in all circumstances because the remaining trees will grow to the same size and quality with or without the additional trees.

While interdependency between the trees of a homogenous stand are necessary for thinnings to be profitable, not all forms of interdependencies are relevant. In case trees improve each other's value growth, they are growing
complementarily, i.e., $pq_{t_k}^j > 0$. This form of interdependence amounts to the
same results regarding the relevance of thinnings as the absence of interdependencies. If the timber value of the remaining trees is impaired by a harvest of a tree prior to the rotation age, it always pays to hold the tree until
the rotation age provided all trees grow equally. Therefore, only the negative
relationship between the postponement of a thinning and the timber value
of a remaining tree offers an argument for thinnings. With reference to the
timber volume, this relationship is referred to as competition, which is frequently defined by its harmful effects on the involved individuals (cf. Begon
et al. 1990, p. 197). The exclusive presence of competition, though, might
only be acceptable in pure stands (cf. Section 3.3.2). In mixed stands, on the
other hand, both competitive and complementary growth might be present.

Beyond the threshold of competition in Figure 2.1, thinnings are thus relevant as they might increase the *LEV* in some situations. Beyond the threshold
of mortality, i.e., the solid black curve Figure 2.1, thinnings become obliga-
tory for homogeneous and equivalent forest stands since trees could be dying off which could be removed with profit (cf. Section 3.3.2). The relevant
range of thinnings is thus delimited by the thresholds of competition and
mortality in Figure 2.1. For positive interest rates, thinnings will be conducted somewhere in-between provided condition [4-21] is satisfied for at
least one tree. As the interest converges to zero, condition [4-13] demands
the timber value increase due to thinnings to equal the lost timber increment
which is just satisfied along the threshold of mortality in Figure 2.1.

Though competition is a necessary condition for thinnings to be relevant in
the homogenous stand, it is not sufficient for thinnings to be conducted profitably. On the one hand, the negative competitive effects need to exceed potential positive effects. On the other hand, according to [4-21], they must also
exceed the land rent in relation to the number of trees in the rotation class.
For given land values and tree numbers, the more intense the competition
for resources, i.e., the higher the potential impact of a harvest of a tree on the

timber value of the remaining trees, the more likely thinnings are profitable as the right and side of [4-21] increases.

The potential sources for intensified competition between trees in a stand are various. On the other hand, the intensity of competition is regulated by the initial density, i.e., the number of trees at the beginning of the rotation. High densities, such as natural regeneration might produce, will increase the portion of thinned trees in contrast to comparatively low planting densities as competition between the trees increases (cf. Section 2.1.1). Likewise, the intensity of competition is dependent on the tree species. The effect of a harvest of a shade-tolerant tree on its conspecifics is usually greater than that in stand of shade-intolerant trees due to asymmetric competition pressures in older age classes. As a consequence, more trees are thinned in a shade-tolerant stand with the same number of trees, all other things being equal. Similarly, competition intensities might be site specific. Forest stands at sites of low soil quality or growing in a disadvantageous climate might grow with large spacing between the trees as some scarce resources restrict the applicability of others. In these cases, thinnings might hardly influence the remaining trees. Thus, all trees will be cut at the rotation age provided they are growing homogenously.

The previous examples are illustrated in Figure 4.3 for imaginary courses of both sides of [4-21]. The abscissa represents increasing tree numbers as originated from the initial density. As the number of trees increases, the potential to improve the growth of remaining trees (dashed curves) increases likewise, however not necessarily at an increasing rate. Additionally, more shade-tolerant tree species or sites of higher quality, for instance, might increase this potential as exemplified by the dashed grey curve.

The other factor separating between thinning and rotation trees in the homogenous stand is the land rent in relation to the square of the number of trees in the rotation class, cf. [4-21]. As the number of trees in the stand n increases for constant k, the land rent is spread over more trees. Consequently, the left hand side of [4-21] decreases as the denominator increases. This relation is indicated in Figure 4.3 by the solid lines. Due to the power function in the denominator, the curve decreases at a decreasing rate. The

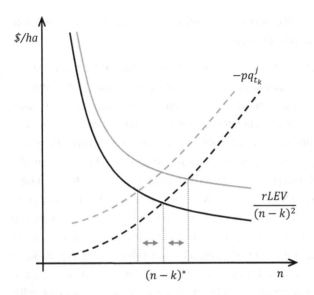

Figure 4.3 The optimal amount of trees $(n - k)^*$ in the rotation class in a homogenous forest stand

value of the curve depends on the land rent. Factors affecting the land rent thus tend to induce shifts as illustrated by the black and grey solid curves in Figure 4.3.

Nevertheless, changes in the investment parameters give rise to adjustments on both sides of [4-21] as the optimal use of the land is modified. For instance, higher timber prices not only increase the land value and the value of the improved timber growth of the remaining trees but also changes the timber growth as the optimal harvest ages are adjusted. Changes in the interest rate will decrease the land value while increasing the interest and simultaneously altering the impact on the remaining trees indirectly through adjusted optimal harvest ages. The same applies to changes in the regeneration cost.

The intersection point between both curves in Figure 4.3 marks the optimal allocation between thinning and rotation trees. All trees at or to the left of the intersection are cut at the rotation age while trees to the right are thinned. If the number of trees in the stand coincides or falls short of the intersection point, thinnings are irrelevant as all trees are necessarily cut at the rotation age. Depending on the actual course of the curve, changes in the investment situation will shift the optimal amount of trees to be thinned. Whenever thinnings are irrelevant in a given situation, the thinning model [3-6] reduces to the Faustmann model [3-2]. Therefore, it might be argued that the Faustmann model refers to all circumstances which lead to equal harvest ages of all trees, such as low competition pressures and high land rents. Density-dependent growth relations, though, are not disregarded implicitly in the Faustmann model [3-2] as these might be built into the timber volume functions although it is impossible for the forest owner to gain control over the density.

In case the rotation class comprises only a single tree, the whole land rent is concentrated on one tree thus increasing the left hand side of [4-21]. The second last tree would then be cut prior to the rotation age only if it exerts an exceptionally high impact on the last tree in order to satisfy [4-21]. In most situations, though, it is not optimal to leave only a single tree until the rotation age because of the unprofitable solitary growth of the last tree. However, if the stand is fairly small and/ or the trees are huge, it might pay to cut the second last tree prior to the rotation age.

Analogously to the determination of the optimal rotation age in Figure 4.1 and of the optimal thinning age in Figure 4.2, respectively, Figure 4.4 illustrates the determination of the optimal rotation age with reference to potential optimal thinning ages for two hypothetical situations A and B, cf. [4-18] and [4-19]. In situation A, the optimal rotation age without any thinnings is located at t_n^{A*} since the rate of cost equals the rate of revenue according to [4-18]. t_n^{A*} without thinnings is equivalent to the optimal Faustmann rotation age T^* in [4-2]. However, Figure 4.4 indicates an incentive to cut parts of the stand earlier as the relative influence on the remaining trees intersects the value growth rate before the optimal rotation age is reached, cf. [4-19]. Therefore, thinnings are relevant in this situation. If thinnings are conducted, the optimal rotation age t_n^{A*} might change and differ from T^*, cf. [4-3].

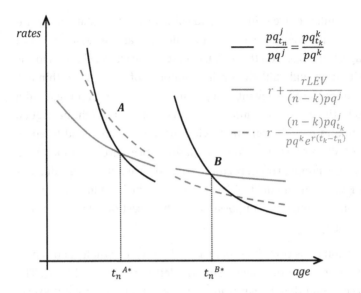

Figure 4.4 Comparison of the relative land rent and the impact rate for two hypothetical situations A and B

In situation B, the optimal rotation age is $t_n{}^{B*}$. It is equivalent to the optimal Faustmann rotation age T^* in [4-2] since thinnings are irrelevant. Although there is a considerable potential to improve the growth of the remaining trees by a thinning as well as there is the opportunity to transform the timber value of the thinning to the next best alternative investment (cf. the dashed curve), it is not enough to sacrifice the growth in value of the kth tree. Hence, the relative costs of a thinning intersect the value growth rate, cf. [4-19], after the optimal rotation age. As this situation conflicts with the model approach, all trees are cut at the rotation age.

4.2.2   In the Heterogeneous Stand

The heterogeneous stand is characterized by potentially diverging timber volumes of equally old trees (cf. Paragraph 2.2). In the heterogeneous stand, condition [4-16] might equally be generalized for any tree k. If the rotation class comprises $n - (k + 1)$ trees, the optimal rotation age is determined by the equality of, cf. [4-3],

$$\frac{\sum_{j=k+1}^n pq_{t_j}^j}{\sum_{j=k+1}^n pq^j} = r + \frac{rLEV}{\sum_{j=k+1}^n pq^j} \qquad [4\text{-}23]$$

while the optimal harvest age of the last tree k considered to be thinned is given by, cf. [4-14],

$$\frac{pq_{t_k}^k}{pq^k} = r - \sum_{j=k+1}^n \frac{pq_{t_k}^j}{pq^k} e^{r(t_k - t_j)}. \qquad [4\text{-}24]$$

This kth tree is added to rotation class of the homogenous stand, i.e., $t_k^* = t_n^*$, if

$$\frac{\sum_{j=k+1}^n pq_{t_j}^j}{\sum_{j=k+1}^n pq^j} - \frac{pq_{t_k}^k}{pq^k} = \frac{rLEV}{\sum_{j=k+1}^n pq^j} + \sum_{j=k+1}^n \frac{pq_{t_k}^j}{pq^k}. \qquad [4\text{-}25]$$

If condition [4-25] holds for all trees in the stand, thinnings are irrelevant as they cannot possibly increase the *LEV* in any investment situation. The same applies if the right hand side of [4-25] outweighs the left hand side since this implies the optimal thinning age to outlast the optimal rotation age which conflicts with the model assumptions, cf. [3-7]. Therefore, for thinnings to be relevant in the heterogeneous stand, it must hold for at least one tree that

$$\frac{\sum_{j=k+1}^n pq_{t_j}^j}{\sum_{j=k+1}^n pq^j} - \frac{pq_{t_k}^k}{pq^k} > \frac{rLEV}{\sum_{j=k+1}^n pq^j} + \sum_{j=k+1}^n \frac{pq_{t_k}^j}{pq^k}. \qquad [4\text{-}26]$$

Accordingly, the difference between the mean value growth rate of the rotation class and the growth rate of the last tree considered to be thinned has to exceed the sum of the land rent in relation to the timber value of the rotation class and the sum of the impacts of a thinning of the kth tree in relation to its timber value.

The right hand side of [4-26] is similar to the conditions for relevant thinnings in the homogeneous stand, cf. [4-20] and [4-21]. Both express the land rent and the impact of a postponed thinning on the remaining trees in relation to the timber value of the stand and to the value of the thinned tree, respectively. The timber values, though, are not necessarily of equal amount in the heterogeneous stand. Therefore, they do not cancel out of [4-26], but, instead, are weighed up against each other. In spite of the differences, the right hand side of [4-26] can be analyzed analogously to [4-21]. With reference to the homogeneous stand, it would be negative for at least one tree if thinnings are relevant and zero or positive for irrelevant thinnings. In contrast to the homogeneous stand, cf. [4-21], the left hand side of [4-26] might be nonzero for equally old trees. If the kth tree, which might be added to the rotation class, is growing in value at a higher, lower or equal rate than the mean of the rotation class, the left hand side is negative, positive or zero. In view of this, the question arises whether all three situations are feasible.

If two homogeneous and equally old stands are compared, of which one is growing in value persistently at a higher rate than the other stand, it cannot be determined which stand is harvested earlier. Although one stand is promising a higher value growth rate, it might be optimal to harvest it before the other stand is cut. This situation may arise since the growth rate is defined only with respect to the timber value, cf. [4-3]. Accordingly, it does not capture all relevant aspects of profitable timber production. This can be seen if the *FPO* theorem [4-2] is rewritten to yield either

$$\frac{pQ_T}{pQ} - \frac{rLEV^s}{pQ} = r$$

$$\Leftrightarrow \quad \frac{pQ_T}{pQ + LEV^s} = r$$

$$\Leftrightarrow \quad \frac{pQ_T}{pQ - C}\frac{e^{rT} - 1}{e^{rT}} = r$$

$$\Leftrightarrow \quad \left(\frac{pQ_T}{pQ}\right)\frac{pQ/LEV}{pQ/LEV + 1} = r.$$

[4-27]

In each case, only adjusted growth rates are relevant for the determination of the optimal rotation age T^*. If, for instance, comparatively high growth rates are combined with equally low regeneration costs, it might pay to cut the more vigorously growing stand sooner. By inference, even when stands are growing at equal rates, they are not necessarily cut at the same age. These aspects are the consequence of the dependence of the income stream of profitable timber production on both timber and land value.

If the two stands are not separated but intermingled as two tree classes in one stand, and if this stand can only be regenerated after both classes are cut, the optimal rotation age would be determined by the *FPO* theorem adjusted for the second stand, i.e.,

$$\frac{pQ_T^T}{pQ^T} = r + \frac{rLEV^+}{pQ^T}, \qquad\qquad [4\text{-}28]$$

where T is the rotation age, Q^T is the timber volume of the class cut at the rotation age, and LEV^+ is the LEV of the merged stand, cf. [4-3]. The optimal harvest age of the second class, however, would be determined according to the optimal thinning condition in Section 4.1.2. If, furthermore, it is assumed that the harvest of one class of the stand leaves the other class unaffected, the optimal thinning age is determined by

$$\frac{pQ_t^t}{pQ^t} = r, \qquad\qquad [4\text{-}29]$$

where t is the thinning age, and Q^t is the timber volume of the thinning class of the stand. Condition [4-29] is equivalent to the solution of the one-rotation or duration problem (cf. Hirshleifer 1970, p. 82ff).

In this setting, the relative land rent, which had to be borne by the thinning class before the merging, has been added to the land rent of the rotation class in [4-30]. Since the opportunity to regenerate the thinning class after it is been harvested has vanished, the opportunity cost of postponing the harvest decrease which in turn increases the optimal harvest age. Conversely, the additional cost comprised in the land value in [4-30] decreases the optimal harvest age of the rotation class in contrast to its separated harvest age.

For equal value growth rates, the optimal harvest age of the thinning class would exceed the optimal harvest age of the rotation class in the same way as the optimal duration age exceeds the optimal rotation age, *ceteris paribus* (Hirshleifer 1970, p. 86; Samuelson 1976, p. 481; Johansson and Löfgren 1985, p. 81). As this conflicts with the assumption that regeneration is linked with the rotation age, both classes would be cut at the rotation age simultaneously. Equally, both are cut at the same age if the differences in the value growth rates between both classes are small enough to be exceeded or offset by the relative land rent in [4-30]. However, if the differences in the value growth rates exceed the relative land rent, i.e., if

$$\frac{pQ_T^T}{pQ^T} - \frac{pQ_t^t}{pQ^t} > \frac{rLEV^+}{pQ^T}, \qquad\qquad [4\text{-}30]$$

the optimal harvest ages in the combined stand will differ. Yet, that class with the higher growth rate will always be cut at the rotation age regardless of the specific investment situation. If the left hand side is negative, [4-30] cannot be satisfied since the right hand side is necessarily positive (cf. Section 3.3.2).

Figure 4.5 illustrates the relationship between two persistently diverging growth rates of two homogenous growing classes of trees or of two single trees, respectively. Depending on each harvest age, the differences between both rates might be positive, negative, or zero. It is zero only if the less vigorously growing class is cut prior; e.g., at the age of t^1 in Figure 4.5 if t^2 is the harvest age of the second class. In case the harvest age is less than t^1 for a given t^2, the difference between both rates as specified in [4-30] is negative since all points above A are of higher value than C in Figure 4.5. Likewise, if the harvest age of the less vigorously growing class is higher than t^1 and less or equal to the given age t^2, the difference between the rates is positive. Therefore, if t^2 is the optimal rotation age in a stand of two independently growing classes, the optimal thinning age will exceed t^1 since the left hand side of [4-30] is only then positive. Invariably, though, the less vigorously

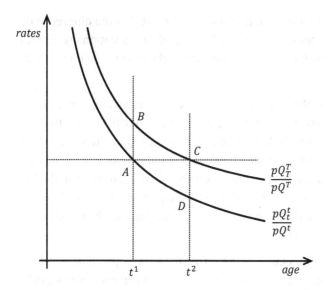

Figure 4.5 An example of the two diverging value growth rates

growing class will not be harvested after the other class is cut as all points above B or C necessarily exceed A or D.

Condition [4-30] can easily be generalized for the heterogonous stand where all trees might grow differently at equal ages. If the harvest ages of each tree are independent of each other, condition [4-30] is analogous to [4-26] without interdependencies, explicitly,

$$\frac{\sum_{j=k+1}^{n} pq_{t_j}^{j}}{\sum_{j=k+1}^{n} pq^{j}} - \frac{pq_{t_k}^{k}}{pq^{k}} > \frac{rLEV}{\sum_{j=k+1}^{n} pq^{j}}. \qquad [4\text{-}31]$$

As a result, and in contrast to the homogenous stand, trees might be harvested prior to the rotation age in the heterogeneous stand if their growth rate is less than the mean growth rate of the trees cut at the rotation age whereas the difference simultaneously exceeds the relative land rent. Or *vice versa*, it is necessary for thinnings to be relevant in a heterogeneous stand of independently growing trees that some trees grow in value at a lower rate than the trees at the rotation age, and it is sufficient that the difference between the growth rates exceeds the relative land rent additionally. Thus, it might be profitable to harvest trees even within the range of solitary growth in Figure 2.1.

As a consequence for the heterogeneous stand, the opportunity arises to cut trees with unsatisfactorily low value growth rates prior to the rotation age. However, these trees will be harvested at a higher age than the same trees in a homogenous stand. By inference, trees with equal growth rates will be cut at the same age if the harvest ages are independent of each other. In the heterogeneous stand, though, trees might share equal growth rates however growing differently. In this case, both the timber value and the increment of one tree are multiples of the other tree.

The preceding analysis remains valid if the impact of a postponement of a thinning on the rotation class is positive since, then, the right hand side of [4-26] is necessarily positive as well. If this impact is accretive to the remaining timber values, or if these positive impacts outweigh or compensate the negative effects, it remains unprofitable to cut more vigorously growing trees previously. For positive impacts, the relevant range for thinnings is shortened, *ceteris paribus*, as the difference between the growth rates must exceed the sum of the relative land rent and the value of the thinning impacts.

In the presence of competition, the impact of a postponement of the harvest reduces the timber volume of the remaining trees. If the overall impact of a postponement is negative, the right hand side of [4-26] might be positive, zero, or negative depending on the investment situation. If the thinning impact is negative but less than or equal to the relative land rent, the right hand side is positive. In this case, the preceding analysis can be applied analogously. Nevertheless, the relevant range for thinnings is extended as the claim to the difference between the growth rates is reduced; i.e., even small differences might justify thinnings. Moreover, only less vigorously growing trees will be cut previously.

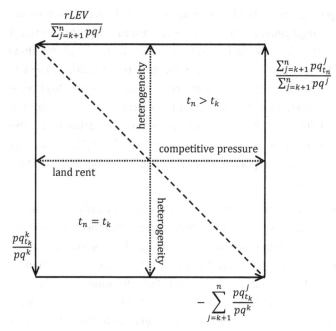

Figure 4.6 The relevant range of thinnings

If, however, the impact of a postponement of a thinning is negative and out-weighs the relative land rent thus turning the right hand side of [4-26] negative, it might pay to cut more vigorously growing trees sooner as the left hand side might become negative while simultaneously satisfying inequality [4-26]. Under these conditions, the value increase of the less vigorously growing trees due to a thinning might justify the sacrifice of the comparatively high value growth rates of the thinned trees. Accordingly, it might pay to cut the superior growing trees of the stand sooner. Under these conditions, thinnings are only relevant when a comparatively strong competition pressure is prevailing.

The preceding analysis is summarized in Figure 4.6. The relevant range of thinnings, as determined by condition [4-26], is pictured by the dashed line in Figure 4.6. Each axis framing the figure denotes increasing rates of one term of [4-26] with equal scales. The axes denoting terms on the same side of [4-26] are opposing each other in opposite direction. Hence, the horizontal arrow within the range generated by the axes represents all points for which the left hand side of [4-26] is zero, i.e., homogeneous stands. All points above and below denote heterogeneous stands as the growth rates differ. The vertical arrow represents all points where the right hand side of [4-26] is zero. The directions of the arrows are marked with the predominant characteristic. For all combinations of the rates on right of the separation line, some trees will be harvested prior to the rotation age, i.e., $t_n^* > t_k^*$. All combinations on or to the left of the separation line will result in an equal harvest age of all trees, i.e., $t_n^* = t_k^*$.

In summary, for thinnings to be relevant, two necessary conditions must either or both be satisfied: heterogeneity concerning differences in the value growth rates and competitive pressure concerning the negative impact on the timber value of remaining trees. If both conditions are not satisfied for any tree in the stand, thinnings are irrelevant. This situation prevails in homogeneous stands of solitarily growing trees. However, in order to be sufficient, both necessary conditions must be set in relation to the relative land rent. In the homogenous stand (horizontal double arrow in Figure 4.6), the existence of competitive pressure alone is not enough to induce thinnings, but must outweigh the relative land rent, which is accomplished to the right of the origin. In situations in which the relative land rent outweighs the competitive pressure (left half of Figure 4.6) only an increasing heterogeneity might induce thinnings. On the other hand, heterogeneity is double-sided since either the more or the less vigorously growing class or tree can be thinned. In case the more vigorously growing class or tree is cut previously (lower half of Figure 4.6), the competitive pressure must outweigh the relative land rent for thinnings to be profitable.

4.3   Optimal Thinning Regime

The optimal thinning regime is typically referred to as the program of the series of thinnings throughout a rotation period (cf. Nyland 2002, p. 448). Since the thinning model [3-6] is maximized with respect to the harvest ages of each tree, these concepts are derived from the individual harvest ages of the trees of the stand. In this way, the thinning regime describes the relationship between the optimal thinning ages. Therefore, whenever thinnings are relevant, the question arises in which specific way thinnings will be conducted most profitably.

Basically, the optimal thinning regime expresses the temporal order of the optimal thinning ages. Accordingly, trees might either be thinned at the same or at successive ages. This distinction gives rise to two fundamental questions: under which conditions are thinnings intensified such that some trees share equal optimal harvest ages (Sections 4.3.1 and 4.3.2)? And which trees have to be cut first, or subsequently respectively (Sections 4.3.3 and 4.3.4)?

4.3.1   Optimal Thinning Intensity

The thinning intensity might be interpreted as the number of trees removed at the same age. In terms of the simultaneous equation system of the necessary condition [3-13] - [3-15], the thinning intensity thus demands the satisfaction of two or more optimal thinning age conditions at the same age. Since the index of the summation in the thinning model [3-6] indicates the temporal order of harvests (cf. Paragraph 3.1), the optimal thinning conditions for two successively considered trees k and $k + 1$ can be written as, cf. [4-14],

$$\frac{pq_{t_k}^k}{pq^k} = r - \sum_{j=k+1}^{n} \frac{pq_{t_k}^j}{pq^k} e^{r(t_k - t_j)} \qquad [4\text{-}32]$$

$$\frac{pq_{t_{k+1}}^{k+1}}{pq^{k+1}} = r - \sum_{j=k+2}^{n} \frac{pq_{t_{k+1}}^j}{pq^{k+1}} e^{r(t_{k+1} - t_j)}. \qquad [4\text{-}33]$$

Since in a maximum all conditions are satisfied simultaneously, [4-32] and [4-33] can be subtracted resulting in

$$\frac{pq_{t_{k+1}}^{k+1}}{pq^{k+1}} - \frac{pq_{t_k}^k}{pq^k} = \sum_{j=k+1}^{n} \frac{pq_{t_k}^j}{pq^k} e^{r(t_k - t_j)} - \sum_{j=k+2}^{n} \frac{pq_{t_{k+1}}^j}{pq^{k+1}} e^{r(t_{k+1} - t_j)}. \qquad [4\text{-}34]$$

If condition [4-34] is satisfied for $t_{k+1}{}^* = t_k{}^*$, both trees k and $k + 1$ are harvested at the same age, which might be equivalent to a more intense thinning. In order to analyze the consequences of this proposition, condition [4-34] is viewed from the two perspectives of the homogenous and heterogeneous stand.

In a homogenous stand, all trees of the same age grow equally (cf. Paragraph 2.1). Therefore, if both optimal harvest ages coincide in the homogeneous stand, i.e., if $t_{k+1}{}^* = t_k{}^*$, both trees share the same value growth rate as $q^{k+1} = q^k$ and $q_{t_{k+1}}^{k+1} = q_{t_k}^k$. In this case, the left hand side of [4-34] is zero thus reducing the expression to

$$\sum_{j=k+1}^{n} \frac{pq_{t_k}^j}{pq^k} e^{r(t_k - t_j)} = \sum_{j=k+2}^{n} \frac{pq_{t_{k+1}}^j}{pq^{k+1}} e^{r(t_k - t_j)}. \qquad [4\text{-}35]$$

Rewriting [4-35] while considering again that $q^{k+1} = q^k$ and $q_{t_{k+1}}^{k+1} = q_{t_k}^k$ for $t_{k+1}{}^* = t_k{}^*$ gives

$$\frac{pq_{t_k}^{k+1}}{pq^k} + \sum_{j=k+2}^{n} \frac{pq_{t_k}^j}{pq^k} = \sum_{j=k+2}^{n} \frac{pq_{t_{k+1}}^j}{pq^{k+1}}$$

$$\Leftrightarrow \quad pq_{t_k}^{k+1} = 0 \qquad\qquad\qquad\qquad\qquad [4\text{-}36]$$

as the condition for a more intense thinning in the homogeneous stand.

According to condition [4-36], a thinning is conducted more intensively in a homogenous stand if the postponement of the thinning does not influence the value of the next tree to be thinned, i.e., if the kth and $k + 1$st tree grow independently of each other. The thinning will be even more intense if condition [4-36] applies to more trees, i.e., $k + 2$, etc. If the condition is never satisfied during the rotation period, the thinning intensity is an empty concept in this setting as all trees thinned are cut at different ages necessarily. This situation occurs when the harvest of any tree in the stand influences all remaining trees. In this case, it always pays to postpone the harvest of the next tree to be cut in order to earn the additional increment generated by the thinning. If, on the other hand, [4-36] is valid for all trees, all of them are cut at the same age, i.e. the rotation age (cf. Section 4.2.1), since the stand is then growing solitarily.

In a heterogeneous stand, the growth rates of two equally old trees might diverge since either or both the timber increment and/ or the timber volume might differ (cf. Paragraph 2.2). As a consequence, the growth rates do not necessarily cancel out of condition [4-34]. Rearranging [4-34] for $t_{k+1}{}^* = t_k{}^*$ with reference to [4-36] yields

$$\frac{pq_{t_{k+1}}^{k+1}}{pq^{k+1}} - \frac{pq_{t_k}^{k}}{pq^{k}} = \frac{pq_{t_k}^{k+1}}{pq^{k}} + \sum_{j=k+2}^{n} \left(\frac{pq_{t_k}^{j}}{pq^{k}} - \frac{pq_{t_{k+1}}^{j}}{pq^{k+1}} \right) e^{r(t_k - t_j)}. \qquad [4\text{-}37]$$

Both trees k and $k + 1$ are harvested at the same age if condition [4-37] holds. The left hand side gives the difference between the value growth rates. It is positive if the second tree to be considered, i.e., $k + 1$, is growing in value at a higher rate than the kth tree, negative in the opposite case and zero for equal growth rates. The first term on the right hand side is the change in the timber value due to a postponement of the thinning of the kth tree in relation to its timber value, which was left in condition [4-36] for the homogeneous stand. The second term on the right hand side gives the present value of the differences between the influences of each tree on the remaining trees in relation to their timber value.

If the trees of a heterogeneous stand grow solitarily, the right hand side of [4-37] is zero as there is no opportunity to influence the timber value of the remaining trees. In this case, the condition for a more intense thinning would demand the growth rates to be of equal magnitude. In a heterogeneous stand, this situation might arise either when some trees are growing homogeneously or when the value increment of some trees is proportionally higher in relation to the employed timber value, e.g., if both the increment and the timber value are twice as high. If thinnings are relevant, those trees will be thinned at the same age.

In contrast, when all trees are distinguished by different value growth rates in a stand of solitarily growing trees, no two trees will be cut at the same age. Therefore, thinnings will not be more intense than the timber volume of one tree. In a homogeneous stand of solitarily growing trees, the difference between the growth rates of equally old trees is always zero. Therefore, the condition for a more intense thinning is satisfied for all trees, i.e., all trees will be cut same aged. However, this will be the optimal rotation age as analyzed in Section 4.2.1.

In a heterogeneous and competitive forest stand, the right hand side of [4-37] might be positive, negative or zero. The second term on the right hand side is zero in the homogenous stand since the equality of the growth of equally old trees implies equal impacts on the remaining trees. Therefore, only the influence between the considered trees remains as the relevant aspect, cf. [4-36]. This might be nonzero in a competitive stand. In the heterogeneous stand, the second term on the right hand side of [4-37] might equally be zero when the impact rates on the remaining trees are of the same magnitude. However, if the impact rates diverge, they might be just as high as to compensate the impact between the considered trees given by the first term on the right hand side. In this case, the right hand side might be zero in the competitive stand.

If the impact rates of both trees k and $k + 1$ on all other trees are equal, i.e., if the second term on the right hand side of [4-37] is zero, and if the harvest ages of both trees are interdependent, i.e., if the first term on the right hand side is nonzero, both trees will only be cut at the same age if their value growth rates differ. If the trees compete in value, i.e., if $pq_{t_k}^{k+1} < 0$, the more vigorously growing tree will be cut sooner, and *vice versa* for $pq_{t_k}^{k+1} > 0$.

Severe thinnings might lead to subsequent phases of solitary growth. As already indicated in the preceding Paragraph 4.2, these are ruled out for optimal thinning ages. Solitary growth implies that the interdependencies between the trees vanish. However, since they have been existent before, condition [4-36] will not hold for all intensities in the homogeneous stand. If the interdependencies are positive or if condition [4-36] applies to all trees, thinnings are irrelevant as all trees will be cut at the rotation age. In the heterogeneous stand, trees with differing growth rates will be only cut at the same age in the presence of interdependencies. As the interdependencies decrease towards solitary growth, some trees will be cut beforehand.

In summary, the optimal thinning intensity is determined by the equality of both sides of [4-37]. Accordingly, the reasons to thin a stand more intensively are given by equal and independent or unequal and interdependent growth. Trees with equal growth rates are only cut at the same age if they are growing independently of each other, or if the difference in the additional impacts on all remaining trees compensates for the loss of an additional increment on the other thinned trees. Trees with differing growth rates might be cut at the same age if the loss of value growth is compensated for by an additional increment on the remaining trees. The more vigorously growing tree is thinned together with the less vigorously growing tree because its impact on the remaining trees is proportionally greater. And the less vigorously growing tree is harvested together with the more vigorously growing tree as it hardly influences the remaining trees.

Other characterizations of the thinning intensity focus on different criteria. For instance, the intensity often refers to the timber volume or basal area removed. If expressed as a percentage, both approaches denote equal intensities in the homogeneous stand since all trees comprise equal timber volumes at equal ages. In the heterogeneous stand, the approaches might diverge as the percentage might either refer to comparatively many and thin or to comparatively few but thick trees. If the intensity refers to an absolute amount, thinnings are intensified for equal numbers of trees harvested at each age in the homogeneous stand while the number of trees in a heterogeneous stand might vary.

4.3.2   Optimal Thinning Frequency

In analogy to the interpretation of the thinning intensity from the preceding paragraph, the thinning frequency can be understood as the frequency of different thinning ages in a stand. In this way, the thinning frequency is the mirror image of the thinning intensity. If a stand is thinned more intensively, i.e., if more tree share the same optimal thinning age, the stand is thinned less frequently, and *vice versa*. Therefore, the more often the condition for a more intense thinning [4-36], or [4-37] respectively, is not satisfied, the more frequently the stand is thinned.

Viewed from this angle, the thinning frequency is the number of thinnings occurring during the rotation age. However, there are often other notions which are related to the frequency of the thinning. This might be the thinning interval as the time period between two thinnings and, in the heterogeneous stand, the type or method of thinning as the order of harvest of differently growing trees. Both aspects will be treated in the following sections. Here, only the thinning frequency is analyzed.

In the homogenous stand, thinnings are conducted more frequently if condition [4-36] does not hold, i.e., if the postponement of the harvest of any tree influences the value of the next tree to be cut. However, in the case of a positive interrelation, the harvest of the kth tree will be postponed necessarily at least to the harvest age of the $k + 1$st tree. Since the trees are growing homogeneously, only positive impacts prevail, and all trees will therefore be cut at the rotation age, as analyzed in Section 4.2.1. In view of this, the condition for a more frequent thinning in the homogenous stand is, cf. [4-36],

$$pq_{t_k}^{k+1} < 0. \qquad\qquad [4\text{-}38]$$

If condition [4-38] is satisfied, the kth tree will be cut prior to the $k + 1$st tree. If [4-38] holds for all trees in the stand, each tree to be thinned will be harvested at a different age. In this case, the thinning frequency might be viewed as an empty concept as the stand will be thinned as frequent as possible, i.e., as numerous as the trees to be thinned. If condition [4-38] is not

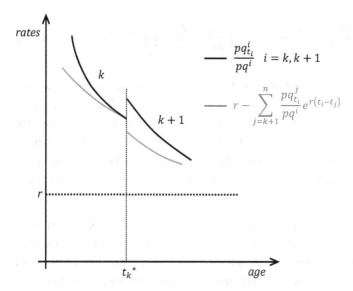

Figure 4.7 Two deviating optimal thinning ages in the homogeneous stand

satisfied for any tree, all trees will be cut at the same age, namely, the optimal rotation age.

Figure 4.7 illustrates the divergence between two optimal thinning ages in the homogeneous stand. Since Figure 4.7 displays a homogeneous stand, all living trees share the same growth rates at equal ages. The kth tree is cut at t_k^* since the value growth rate (black curve) equals the relative impact on the remaining trees in addition to the rate of interest (grey curve) at this age. On the contrary, the $k + 1$st tree will be cut at a later age since its value growth rate is above its relative impact on the remaining trees. The jump in the growth rate is induced by the additional increment by virtue of the removal of the kth tree. Without this impact on the $k + 1$st tree, both trees will be cut at the same age.

Analogously, in the heterogeneous stand, trees are cut at different ages if condition [4-37] does not hold. Since the marginal revenues of holding the $k + 1$st tree have to outweigh its associated marginal costs if the tree has not reached its optimal harvest age yet, it must hold that

$$\frac{pq_{t_{k+1}}^{k+1}}{pq^{k+1}} - \frac{pq_{t_k}^{k}}{pq^{k}} > \frac{pq_{t_k}^{k+1}}{pq^{k}} + \sum_{j=k+2}^{n} \left(\frac{pq_{t_k}^{j}}{pq^{k}} - \frac{pq_{t_{k+1}}^{j}}{pq^{k+1}} \right) e^{r(t_k - t_j)}. \qquad [4\text{-}39]$$

Otherwise, the order of harvests denoted in the summation index is confused. If condition [4-39] is satisfied, then the kth tree is harvested prior to the $k + 1$st tree. Accordingly, in order to be harvested at a higher age, the difference in the value growth rate of the $k + 1$st tree to the value growth rate of the kth tree must outweigh both the impact rate between both trees as well as the difference in the impact rates of each tree on all remaining trees.

As analyzed in the preceding Section 4.3.1, the right hand side of [4-39] might be positive, negative or zero in the heterogeneous stand. Consequently, thinnings are conducted more frequently if trees growing at equal rates either are in competition or differ in their impact on the remaining trees. Likewise, more thinnings will be conducted if the differences in the value growth of trees are not compensated by positive impacts of the less vigorously growing trees or by high negative impacts of the more vigorously growing trees.

Figure 4.8 illustrates the occurrence of deviating optimal thinning ages in a heterogeneous stand. At the first optimal thinning age $t_k{}^*$, growth rates and relative impacts on the remaining trees of the other two trees $k + 1$ and $k + 2$ diverge. The higher growth rates might be reinforced by the harvest of the kth tree, or independent of it. At the second optimal thinning age $t_{k+1}{}^*$, the more vigorously growing tree is harvested as its high impact rate compensates for its high growth rate. The other tree is not influenced by the thinning in this illustration as its value growth rate remains unchanged. However, since its impact on the remaining trees is comparatively low, thinnings are conducted more frequently by postponing the harvest of the $k + 2$nd tree.

As a result, the thinning frequency is determined by the interaction of heterogeneous growth and negative interdependencies. Heterogeneous stands are thinned frequently unless strong interdependencies compensate the differences in the growth rates. The more independently trees are growing and the less heterogeneously, the less frequently the stand is thinned. Trees to be

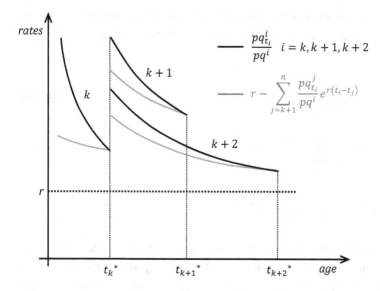

Figure 4.8 Deviating optimal thinning ages in a heterogeneous stand

thinned in overall heterogeneous and solitarily growing stands as well as trees to be thinned in homogenous and overall competitively growing stands are all harvested at different ages.

4.3.3   Optimal Thinning Interval

The optimal thinning interval denotes the time elapsed between two thinnings. Naturally, this concept is only meaningful if thinnings are more or less frequent such that some optimal thinning ages diverge. In this qualitative approach, the length of the thinning interval can only be assessed in relation to other intervals within the same rotation period. In this way, the thinning intervals might be decreasing, increasing or constant over the stand age if the intervals between two thinnings become shorter, longer or remain unchanged.

Basically, the sequence of the thinning intervals is determined by two successive thinning intervals which are in turn determined by three harvest ages, i.e.,

$$\frac{pq_{t_k}^k}{pq^k} = r - \sum_{j=k+1}^n \frac{pq_{t_k}^j}{pq^k} e^{r(t_k-t_j)} \qquad\qquad [4\text{-}40]$$

$$\frac{pq_{t_{k+1}}^{k+1}}{pq^{k+1}} = r - \sum_{j=k+2}^n \frac{pq_{t_{k+1}}^j}{pq^{k+1}} e^{r(t_{k+1}-t_j)} \qquad\qquad [4\text{-}41]$$

$$\frac{pq_{t_{k+2}}^{k+2}}{pq^{k+2}} = r - \sum_{j=k+3}^n \frac{pq_{t_{k+2}}^j}{pq^{k+2}} e^{r(t_{k+2}-t_j)}. \qquad\qquad [4\text{-}42]$$

Denoted as rates, both sides of the equations are strictly monotonously decreasing functions at decreasing rates. The value growth rates on the left hand side are decreasing due to the sufficient condition [3-16] which restricts the attention of the maximization to the domain where the acceleration rates are less than the growth rates (cf. Section 3.3.2). Likewise, the impact rates are monotonously decreasing since the greatest impact on the remaining trees is exerted the earlier the tree is cut while it is constant during the phase of solitary growth. However, since the denominator is necessarily increasing (cf. Section 3.3.2), the impact rate is decreasing even within the solitary range.

At a maximum of the *LEV*, equations [4-40] - [4-42] are satisfied simultaneously. Isolating the rate of interest in all three conditions and equating subsequent harvests yields

$$\frac{pq_{t_{k+1}}^{k+1}}{pq^{k+1}} - \frac{pq_{t_k}^k}{pq^k} = \sum_{j=k+1}^n \frac{pq_{t_k}^j}{pq^k} e^{r(t_k-t_j)} - \sum_{j=k+2}^n \frac{pq_{t_{k+1}}^j}{pq^{k+1}} e^{r(t_{k+1}-t_j)} \qquad [4\text{-}43]$$

$$\frac{pq_{t_{k+2}}^{k+2}}{pq^{k+2}} - \frac{pq_{t_{k+1}}^{k+1}}{pq^{k+1}}$$
$$= \sum_{j=k+2}^n \frac{pq_{t_{k+1}}^j}{pq^{k+1}} e^{r(t_{k+1}-t_j)} - \sum_{j=k+3}^n \frac{pq_{t_{k+2}}^j}{pq^{k+2}} e^{r(t_{k+2}-t_j)}. \qquad [4\text{-}44]$$

Subtracting [4-43] from [4-44] leads to

$$
\left(\frac{pq_{t_{k+2}}^{k+2}}{pq^{k+2}} - \frac{pq_{t_{k+1}}^{k+1}}{pq^{k+1}}\right) - \left(\frac{pq_{t_{k+1}}^{k+1}}{pq^{k+1}} - \frac{pq_{t_k}^{k}}{pq^{k}}\right)
$$

$$
= \left[\sum_{j=k+2}^{n} \frac{pq_{t_{k+1}}^{j}}{pq^{k+1}} e^{r(t_{k+1}-t_j)} - \sum_{j=k+3}^{n} \frac{pq_{t_{k+2}}^{j}}{pq^{k+2}} e^{r(t_{k+2}-t_j)}\right] \qquad [4\text{-}45]
$$

$$
- \left[\sum_{j=k+1}^{n} \frac{pq_{t_k}^{j}}{pq^{k}} e^{r(t_k-t_j)} - \sum_{j=k+2}^{n} \frac{pq_{t_{k+1}}^{j}}{pq^{k+1}} e^{r(t_{k+1}-t_j)}\right].
$$

Accordingly, it must hold that the difference in the differences in the value growth rates equals the differences in the differences of the impacts on the remaining trees.

In the homogeneous stand, the left hand side of [4-45] is necessarily positive if the value growth rates are monotonously decreasing at decreasing rates and if the intervals between the harvest ages are of equal or of decreasing length. This setting, however, does only remain valid as long as the interrelations between the thinned trees do not differ substantially as otherwise the differences might be compensated. By inference, if the left hand side of [4-45] is negative, the thinning interval must be increasing over the stand age. Therefore, if the right hand side is negative necessarily, constant or decreasing thinning intervals can be excluded. The right hand side is negative whenever the impact rate of the $k + 1$st tree is higher than the mean of the kth and the $k + 2$nd tree. This applies either or both if the opportunity to influence the remaining trees is increasingly reduced due to already conducted thinnings and/ or if the impact rates are decreasing substantially.

Figure 4.9 illustrates this argument. It displays a situation where the optimal thinning interval is constant as determined by the intersection points of the value growth rates (black curves) and the impact rates in addition to the interest rates (grey curves). As a result, both sides of [4-45] are positive. The optimal thinning interval tends to decrease if the thinning induced reductions of already conducted thinnings, i.e., the gaps between the grey curves,

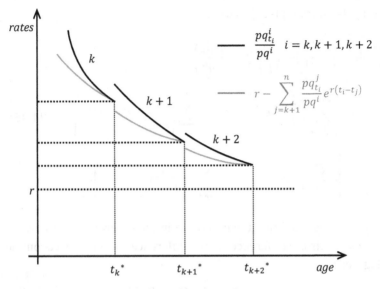

Figure 4.9 A constant optimal thinning interval

shrink. The same applies if the slopes of the impact rates decrease after each conducted thinning. In either way, both the value growth rate and the impact rate are decreasing at decreasing rates. Therefore, the comparatively high reactivity of younger stands might be compensated by the comparatively high value growth while the opposite might apply in older stands. In view this approach, the optimal thinning interval cannot be specified in general. Both increasing and decreasing as well as constant thinning intervals are conceivable depending on the specific investment situation.

In the heterogeneous stands, thinnings might be conducted without mutual interdependencies between the trees (cf. Section 4.2.2). In this case, the right hand side of [4-45] is zero. In order to satisfy the equality, the left hand side has to turn zero likewise as is the case when all trees are thinned when their value growth rate equals the rate of interest. If younger stands produce more trees of low value growth as older stands, the thinning interval must increase over the age, and *vice versa*. With competitive pressure, the heterogeneity in the value growth rates might be compensated by the impact rates. Hence, less vigorously growing trees with comparatively low impact rates and more vigorously growing trees with comparatively high impact rates might be cut in increasing, decreasing or constant intervals. For instance, the increasing thinning interval in Figure 4.8 turns into a decreasing interval if the impact rate of the k + 2nd tree has a lower slope, if the impact rate of the kth tree is higher or if value growth rate is lifted by previously conducted thinnings.

4.3.4   Optimal Thinning Method

The optimal method or type of thinning usually denotes the order of harvests of the trees in a stand on the basis of some classification of the trees (cf. e.g. Smith et al. 1997). These classifications might be based on different criteria. For instance, there might be crown, height, diameter or quality criteria. In view of the model approach of this study (cf. Paragraph 3.1), trees might be classified according to their value growth rates as these help determining the optimal harvest ages, cf. [4-3] and [4-14]. Accordingly, trees growing in value at high rates at a specific age denote high tree classes, and trees growing at low value rates low tree classes. Other classification criteria, on the other hand, such as the timber or value increment, might cause ambiguities as different classifications might be applied in the same investment situation. The classification does not refer to specific trees, but the membership to the classes is variable and dependent on the age. Thus, trees might belong to different classes during the rotation age. Typically, all trees of a stand belong to the same class at the very beginning of the rotation period of an even-aged stand as factors of differentiation (cf. Paragraph 2.2) have not been effective yet.

Based on this classification, two different types of thinnings might be distinguished: the harvest of less vigorously growing trees prior to more vigorously growing trees, and the reverse case. In the simplified approach employed hitherto, the development of the tree value is only dependent on the development of the timber volume such that timber value growth rates equal timber volume growth rates (p might be cancelled out of the value growth rates). Therefore, trees are classified in the order of their volume growth rates. In this case, a connection might be set-up to the classical silvicultural thinning methods. Typically, these are defined in terms of the tree classes according to Kraft (1884; cf. Smith et al. 1997, p. 99 ff.). Thereafter, trees are classified with reference to their relative dominance. The latter is determined by crown and position criteria such as crown size, vitality and shape as well as tree height. The higher a tree in relation to the other stand members, and the larger and healthier its crown, the higher is its degree of dominance. High tree classes comprise dominating trees with voluminous, healthy crowns while dominated trees in low tree classes have comparatively small, squeezed or dying back crowns.

The objective of this often termed "natural" (Assmann 1970, p. 83) classification scheme might be interpreted to organize trees in classes of equal "growth energy" (Assmann 1970, p. 84). Naturally, bigger crowns give rise to greater absolute timber increments due to the pipe model theory (cf. Section 2.1.2). However, Assmann (1970 p. 99) gives evidence that the Kraft (1884) classes not only arrange trees according to their timber increments but to their timber volume growth rates (cf. also Magin 1952). In this way, trees of higher classes produce disproportionately more timber volume such that their comparatively higher timber volume is overcompensated. This disproportional production of timber in relation to the employed timber volume follows from the higher efficiency of dominant trees to utilize resources (cf. Boyden et al. 2008; Campoe et al. 2013; Gspaltl et al. 2013).

If the Kraft (1884) classes arrange the trees in the order of their volume growth rates, and if the Kraft (1884) classes serve to define the silvicultural thinning methods, the harvest of less vigorously growing trees prior to more vigorously growing trees might be termed thinning from below (equivalent to low thinnings, cf. Smith et al. 1997, p. 99 f.). The opposite case, when more vigorously growing trees are harvested previously, might refer to either crown or selection thinnings (Smith et al. 1997, p. 102 ff.) specified by the tree classes remaining in the stand. For simplicity, these thinning methods are termed thinnings from above in the following. Depending on the number of trees thinned simultaneously, different grades or intensities of these thinning methods might be distinguished where increasing portions of the upper or lower growth rate distribution are removed.

The allocation of trees to classes implies homogeneity within the classes since these trees are treated equally. Moreover, when the tree classes are assumed to be intensively intermingled on the stand area, the trees within a class are growing more or less independently of each other. If this applies, the whole class is optimally harvested if one of its trees has reached its optimal harvest age (cf. Section 4.2.1). Therefore, removals of parts of a tree class imply either interdependent growth of its trees or heterogeneity within the class. In the former case, the optimal thinning intensity in the homogeneous stand has to be considered while in the latter case the condition for the optimal thinning intensity in the heterogeneous stand additionally applies (cf. Section 4.3.1).

In a homogenous stand, the concept of thinning methods is irrelevant. Since all trees are growing equally at equal ages, there is no distinction between more or less vigorously growing trees. Thinnings are then neither from above nor from below, or both at the same time, depending on the point of view. Consequently, intentions to improve the growth of selected trees are only focused on all remaining trees but not on single trees. Here, all trees serve the same end with equal shares. Although trees might be harvested prior to the rotation age in order to improve the value of remaining trees (cf. Section 4.2.1), the order of harvests or the assignment of index numbers to each tree, respectively, is irrelevant as the stand grows homogeneously afterwards. Put in other words, it is irrelevant which trees are removed since all trees grow and influence other trees equally.

In the heterogeneous stand, by contrast, trees might grow and influence the remaining trees differently. Hence, the opportunity arises to either cut more or less vigorously growing trees previously. However, without mutual interdependencies between the trees, the order of harvest is determined solely by the value growth rates. As already indicated in Section 4.2.2, the optimal thinning age is then determined by

$$\frac{pq_{t_k}^k}{pq^k} = r. \qquad\qquad [4\text{-}46]$$

Since the volume growth rates are strictly monotonically decreasing functions (cf. Section 3.3.2), trees with lower value growth rates are cut prior to more vigorously growing trees necessarily as they intersect the constant rate of interest at an earlier age, provided they outweigh the influence of the relative land rent (cf. Section 4.2.2). This necessary order of the harvest ages due to the growth rates might have encouraged Pressler to demand the harvest of all "lazy or overripe workers" (1865, p. 18, original: 'faule oder überreife Gesellen'), which are not promising to yield a value increment above the interest rate. From this point of view, only thinnings from below are relevant in solitarily growing stands if value and volume growth rates are identical.

When mutual interdependencies between trees are included, the order of harvests according to increasing growth rates is not necessary. Considering the conditions for the divergent optimal harvest ages of two successively thinned trees k and $k + 1$, i.e., [4-32] and [4-33] with $t_{k+1}{}^* > t_k{}^*$ at the age $t_k{}^*$, it must hold at a maximum of the LEV that (this is a reproduction of condition [4-39])

$$\frac{pq_{t_{k+1}}^{k+1}}{pq^{k+1}} - \frac{pq_{t_k}^{k}}{pq^{k}} > \frac{pq_{t_k}^{k+1}}{pq^{k}} + \sum_{j=k+2}^{n} \left(\frac{pq_{t_k}^{j}}{pq^{k}} - \frac{pq_{t_{k+1}}^{j}}{pq^{k+1}} \right) e^{r(t_k - t_j)}. \qquad [4\text{-}47]$$

Accordingly, if $t_{k+1}{}^* > t_k{}^*$, the difference between the value growth rates on the left hand side must outweigh the difference between the relative impacts on the remaining trees on the right hand side.

Applying the basic distinction between thinning from above and below from the beginning of this section, the left hand side of [4-47] is positive if the first tree to be cut, i.e. k, grows at a lower rate, and it is negative if the kth tree grows more vigorously. If thinning from below is more profitable, the right hand side of [4-47] might be either positive, zero or negative. If in this case the more vigorously growing tree exerts no influence on any tree, the left hand side of [4-47] is necessarily negative when only competitive pressure exists, thus satisfying the inequality. The stronger the impact of the more vigorously growing tree on the remaining trees, the more likely the right hand side might turn positive. When the impact in relation to the impact of the less vigorously growing tree is strong enough, the inequality will not hold. Then, both trees are either cut simultaneously or the more vigorously growing tree will be cut previously.

If thinning from above is more profitable, the left hand side of [4-47] is negative. In order to satisfy the condition, the right hand side must turn negative as well. If the impact rates on the remaining trees are similar, this will only apply when the more vigorously growing tree exerts a considerable impact on the less vigorously growing tree. However, if the impact of the more vigorously growing tree on the remaining trees is strong enough, it might outweigh both the difference in the growth rates and the impact of the less vigorously growing tree.

In this approach, different thinning methods might be characterized according to the order of harvests set by the value growth rates as long as the net unit revenues remain unique for all trees (cf. Paragraph 4.4). Table 4-1 provides an overview on the basis of condition [4-47]. In the homogeneous stand, thinnings might be either more or less intense depending on the mutual interdependencies (cf. Sections 4.3.1 and 4.3.2). Since all trees grow equally, the left hand side as well as the second term on the right hand side are necessarily zero. In the heterogeneous stand, thinnings might be conducted either from above or from below depending on the characteristic that more vigorously growing trees are cut subsequently or prior to less vigorously growing trees. In each case, equally growing trees within the heterogeneous stand might be cut simultaneously as indicated by the zero on the left and sides of the equalities.

Within thinnings from below, different grades might be distinguished according to the removal of more or less differently growing trees. Low grades (e.g. grade-A) only remove equally and less vigorously growing trees which might or might not exert influence on each other or the remaining trees. At the same time, more vigorously growing trees are left in the stand, as indicated by the inequality. Higher grades of thinning from below (e.g. grade-B or -C) simultaneously remove more vigorously growing trees but also leave

Table 4-1 Characteristics of different thinning methods in view of the model approach

$$\frac{pq_{t_{k+1}}^{k+1}}{pq^{k+1}} - \frac{pq_{t_k}^{k}}{pq^{k}} \quad \begin{Bmatrix} = \\ > \end{Bmatrix} \quad \frac{pq_{t_k}^{k+1}}{pq^{k}} \quad + \quad \sum_{j=k+2}^{n} \left(\frac{pq_{t_k}^{j}}{pq^{k}} - \frac{pq_{t_{k+1}}^{j}}{pq^{k+1}} \right) e^{r(t_k - t_j)} \qquad [4\text{-}47]$$

	(0)	=	(0)	+	(0)	more intense		homo-geneous
	(0)	>	(−)	+	(0)	less intense		
	(0)	=	(0/−)	+	(0/+)	grade-A		
∧	(+)	>	(0/−)	+	(+)		thinning from below	heterogeneous
	(0)	=	(0/−)	+	(0/+)			
∧	(+)	=	(0/−)	+	(+)	grade-B(C)		
∧	(+)	>	(0/−)	+	(+)			
	(0)	=	(0/−)	+	(0/+)	selection		
∧	(−)	>	(−)	+	(−)	thinning		
	(0)	=	(0/−)	+	(0/+)			
∧	(−)	>	(−)	+	(−)	crown thinning	thinning from above	
∧	(+)	>	(−)	+	(−)			

even more vigorously growing trees in the stand, i.e., an increasing portion of the lower value growth rate distribution is thinned.

Within thinnings from above, different types might be distinguished as well. Selection thinning is often referred to as the removal of the most dominant trees (Smith et al. 1997, p. 107 ff.), which will be the trees with the highest growth rates in the classification applied. For instance, this offers an argument for removing wolf trees in young stands. It implies, however, that inequality [4-47] holds; i.e., the negative relative influence of the removed trees is strong enough in order to offset the differences in the growth rates.

In each case, by adding competitive pressures, the optimal order of harvests becomes ambiguous and not predetermined within the model approach. On the one hand, more vigorously growing trees earn a higher rate of return as less vigorously growing trees. On the other hand, more vigorously growing trees exert a stronger impact on the remaining trees as they release more resources when harvested. Less vigorously growing trees perform comparatively poorly but do hardly influence the remaining trees. Moreover, trees growing at a high rate might have accumulated high timber values thus reducing the relative impact on the remaining trees even with strong absolute impacts.

Basically, two conditions might be distinguished which favor thinnings from above. The higher the negative impact of the more vigorously growing tree on the remaining trees and the less the impact of the less vigorously growing tree on the remaining tree, the more likely thinnings are conducted from above as the right hand side of [4-47] turns negative. Alternatively and/ or additionally, positive impacts of the less vigorously growing tree on the remaining trees likewise increases the likelihood of thinnings from above. In the reverse case, i.e., with high impacts of the less vigorously growing trees and/ or low impacts of the more vigorously growing trees, thinnings are more likely to be conducted from below.

4.4   Volume and Value

The preceding analysis has focused primarily on the timber value and its change as the relevant determinants of the optimal harvest ages (cf. Section 4.1.1). In the general model approach [3-6], the timber value only varies with the timber volume. In this section, the analysis is specified to account for different sources of timber value. This specification includes price differentials for different timber structures (Section 4.4.1) as well as the influence of harvest costs (Section 4.4.2). Moreover, the impact of the regeneration cost on the optimal thinning regime is analyzed (Section 4.4.3). In this context, the approach is extended to include the initial density as an endogenous variable since it directly sets the relevant range for thinnings. Finally, the optimal timber volume as a consequence of the individual harvest ages is analyzed (Section 4.4.4).

4.4.1   Quantity and Quality

The timber value of a stand is the product of the timber volume with the timber price at which it can be sold. Timber volume, however, might be a composite expression which summarizes different qualities of timber as a uniform product. In order to separate both notions, timber volume is referred to as the combination of timber quantity and timber structure. The former is understood as the physical extension of timber. The timber structure, by contrast, is the combination of the quality and the dimension of the timber. Timber quality is typically determined through criteria concerning the inner structure of the timber. In this approach, quality criteria might be anything for which different timber prices are yielded. For instance, these criteria might refer to knots, juvenile wood, fungus infested areas, straightness or to spiral grain. The timber dimension, on the other hand, refers to the outer structure, which is determined by the diameter and length of the timber.

Analogously to the timber volume (cf. Chapter 2), some aspects of the timber structure are also sensitive to the harvest ages in a stand. First of all, the stem diameter of a tree varies with its own harvest age and, potentially, with the harvest ages of its neighboring trees (cf. Section 2.1.2). In the same way, branch diameters as quality criteria respond to changes in the harvest ages. Other aspects of the timber structure, on the other hand, might not be influenced by the harvest ages, such as fungus infested areas. These are irrelevant for a qualitative analysis as they do not influence the optimality of the endogenous variables.

The relevance of these timber characteristics hinges on price differentials. In the presence of a uniform timber price, only the timber volume is relevant for the determination of the timber value. With price differentials, equal timber volumes might represent different timber values due to different timber structures. Therefore, a specific timber price exists for each timber volume with a specific timber structure. From this point of view, different timber prices might be obtained if changes in the timber structure as a consequence of changes in the harvest ages occur. In this way, unit revenues for timber are age-dependent in the face of price differentials. While timber prices are thus exogenously given, unit timber revenues might change endogenously.

The age-dependence of the unit revenues refers to changes in the timber structure. By contrast, unit timber revenues might also be dependent on time as timber prices might change in-between (cf. Section 3.3.1). In the static approach within the "Faustmann laboratory" (Deegen et al. 2011, p. 363 ff.), however, time and age are synchronized such that rate of price change is constant. In this as in most analyses, the rate of price change is assumed to be zero (cf. Section 3.3.1). Analyses with nonzero rates of price changes are provided, e.g., by McConnell et al. (1983), Newman et al. (1985) and Yin and Newman (1995). Nevertheless, the static character of the analysis remains as the rate of price change is constant and known. In dynamic optimization problems, the rate of price change might be irregular.

The relevance of these different sources of timber value is at least well-known since Pressler (1860; 1995). With reference to the change in the timber value, Pressler distinguished between the quantity increment, the quality increment and the price increment (Pressler 1860, p. 174). In their attempt to show that the rates of the different increments, adjusted for the timber and land value and condensed as Pressler's indicator rate, represent the first order condition for both the maximization of the Faustmann model [3-2] as well as for the maximization of the generalized Faustmann model (Chang 1998), Chang and Deegen (2011) were able to separate the different increments analytically. Accordingly, they interpret the stumpage value of a stand as the sum of the timber values of product classes, which are characterized by a particular timber price. The value of a product class is generated as the product of its share of total stand volume and the corresponding timber price (Chang and Deegen 2011, p. 261). In this setting, the timber volume of the stand is a heterogeneous accumulation of different timber structures saleable at different prices, which are all subject to changes over the age. Furthermore, this approach enables to analyze the influence of the planting density on the profitability with the inclusion of quality aspects (Coordes 2013).

In this study, and in contrast to Chang and Deegen (2011), the timber volume of the stand is additionally separated temporally, i.e., different timber volumes might be harvested at different ages. Accordingly, the timber volume of a specific tree q^i is modeled to yield a particular unit revenue p^i. In terms of Chang and Deegen (2011), all trees yielding a specific timber price thus represent a product class. However, trees usually comprise different timber structures which might be sold at different prices (e.g. logs and brushwood). Therefore, the particular unit revenue p^i a tree yields has to be interpreted as the weighted mean of the different prices for each product class which are comprised in a tree. It is thus assumed that the tree is the inseparable unit of the forest stand. Though this assumption might be less severe for the application of the model to problems of timber production, it might be important for the production of products other than timber (cf. Li and Löfgren 2000). In these cases, however, in which the separability of a tree is relevant, the problem might be solved analogously to the repeal of the inseparability of the forest stand (cf. Paragraph 3.1).

In the preceding analysis, only the total change in the timber value has been addressed. This total change might now be separated into the value change due to changes in the timber volume of each tree q^i and due to changes in the timber structure. The latter becomes relevant through changes in the unit revenues p^i, which can be obtained for the modified timber structure. Since the timber structure potentially varies with the harvest ages, it is assumed that

$$p^i = p^i(t_1, \dots, t_i), \tag{4-48}$$

i.e., the unit revenue is potentially dependent on all previously conducted harvests as well as on the harvest of the specific tree itself in the same way as the timber volume, cf. [3-5].

The direction of change, though, is dependent on the specific situation. For instance, timber of larger stem diameters usually yields higher timber prices. In this case, it holds that

$$\frac{\partial p^i}{\partial t_i} > 0; \quad \frac{\partial p^i}{\partial t_j} \leq 0 \ with \ j \in \{1, \dots, i - 1\} \qquad \qquad [4\text{-}49]$$

since postponements of the harvest age of a tree increase the stem diameter while postponements of previously conducted thinnings might reduce the diameter (cf. Section 3.3.2). Nevertheless, the unit revenue due to higher diameters might also be reversed as is typically observed for very large diameters due to their problematic processing. Just the opposite of [4-49] might also apply to changes in the timber quality. For instance, larger branch diameters often result in lower timber quality. Since branch diameters respond to changes in the harvest ages in the same way as stem diameters, the unit revenue might change in the opposite direction of [4-49]. In either case, the existence of unit revenue impacts implies volume impacts since there is no unit revenue impact without a volume impact.

With the description in [4-48], the timber value pq^i is specified as

$$pq^i(z) = p^i q^i(z) = p^i(z) q^i(z), \qquad \qquad [4\text{-}50]$$

where z is a vector comprising all relevant variables, i.e. in this case, the harvest ages $t_1, \dots, t_i$. As a consequence, the change in the timber value due to a change in the variables can be separated into changes in the timber volume and the unit revenue, respectively, since

$$\frac{\partial [p^i(z) q^i(z)]}{\partial z_i} = [p^i q^i]_{z_i} = p^i_{z_i} q^i + p^i q^i_{z_i}. \qquad \qquad [4\text{-}51]$$

For instance, the revenues on the left hand side of the condition for the optimal rotation age [4-1] is divided to give the value increment due to the change in the unit revenue that can be obtained for the changing timber structure and the change in the quantity of the timber volume.

4.4.1.1   Optimal Thinning Ages with Price Differentials

The separation of the value increment into structural and voluminous changes in the timber allows specifying the propositions concerning the optimal thinning regime in the preceding Paragraphs 4.1 - 4.3 since thinnings are often justified in order to increase the quality of the remaining trees. While thinnings might not increase the timber quantity at the rotation age, i.e., in particular, every thinning with rotation ages located before the mortality threshold in Figure 2.1, thinnings might increase the timber value by virtue of higher unit revenues which can be obtained at the rotation age. Loss of timber volume at the rotation age as a result of thinnings might increase the *LEV* as long as the sum of the unit revenue and the volume increment, cf. [4-51], remains positive.

With the separation in [4-51], the maximum condition for the optimal thinning age [4-14] can be rewritten as

$$\frac{p_{t_k}^k}{p^k} + \frac{q_{t_k}^k}{q^k} = r - \sum_{j=k+1}^{n} \left(\frac{p_{t_k}^j q^j}{p^k q^k} + \frac{p^j q_{t_k}^j}{p^k q^k} \right) e^{r(t_k - t_j)}. \qquad [4\text{-}52]$$

Here, the value growth rates are separated into unit revenue and volume growth rates. For constant unit timber revenues, both the unit revenue changes and the timber prices as a whole can be eliminated in [4-52]. In this case, only the impact on the timber volume remains relevant such that a gain in timber volume now is balanced with an additional timber volume at future harvest ages. With price differentials, additional incentives for thinnings might arise.

If the unit revenue is positively correlated with the stem diameter, for instance as the relevant criterion of the timber dimension, both unit revenue changes on either side of [4-52] are positive since either the potentially thinned tree grows thicker or the remaining trees increase their diameter growth. Hence, both the marginal revenues and the marginal costs increase. If the unit revenue increases at a decreasing rate over rising diameters, the price increase of younger and thinner trees is greater than the one of older and thicker trees for equal increases in the increment. If, then, the diameter

growth of younger trees is greater than the responses of older trees, the optimal thinning age tends to increase as the marginal revenues increase more
than the marginal costs. By contrast, the relevance of thinnings in the homogeneous stand expands in this setting since either the *LEV* in [4-20] decreases due to lower unit revenues for thinned trees or both sides of [4-20]
increase due to higher unit revenues for trees cut at the rotation age. In this
way, larger price differentials between diameters favor thinnings but only
relatively late thinnings.

The same applies to linearly increasing unit revenues since the future unit
revenue increase is discounted and the reaction of the older tree might be
less intense (cf. Section 2.1.4). For progressively increasing unit timber revenues, on the other hand, as well as for more vigorously growing older trees
and for heterogeneous stands, earlier thinnings might become relevant
when the increase in the future timber value justifies the comparatively
small loss of the thinned tree. If the unit revenue increase of the remaining
trees is assumed high enough, already small impacts will justify significant
sacrifices of the thinned trees. At the same time, however, the relevance of
thinnings increases as the right hand side of [4-26] decreases.

With the introduction of quality criteria, additional incentives arise which
are influenced by thinnings. If the unit revenue regarding the quality is negatively correlated with the diameter, for instance due to larger branch diameters, a thinning causes a unit revenue decrease of the remaining trees. At
the same time, its own unit revenue is prevented from decreasing due to a
lower quality. Together, the optimal thinning age thus decreases for linearly
or regressively decreasing unit revenues over the diameter necessarily since
the marginal costs are reduced more sharply than the marginal revenues.
The relevance of thinnings, however, is diminished as condition [4-20] is less
likely to be satisfied due to the decrease of the right hand side and the increase of the left hand side. The same applies to progressively decreasing
unit revenues although the optimal thinning age might then increase.

This effect might be more or less irrelevant for living branches in many cases
since the decisive branches for the timber value are usually those at the

lower parts of the stem as these determine the timber structure of the greatest part of the timber volume. Often, these lower branches are dead when
thinnings are conducted and might thus not increase their growth. In other
cases, however, the thicker branches due to thinnings might be important;
for instance, in young, very dense stands where early thinnings exert a substantial impact on both stem and branch diameters of the remaining trees.
Viewed from a different perspective, however, thinnings might also increase
the growth of branches which have not been growing yet. Those parts of the
stem with dead branches might grow water shoots in the sequence of thinnings due to the increase of available resources.

Both positive and negative interdependencies often arise simultaneously.
For instance, considering a more or less homogenous and competitive stand
of a shade-tolerant tree species like European beech (*Fagus sylvatica* L.) or
sugar maple (*Acer saccharum* Marsh.), the trees are competing for resources
thus reducing each other's timber growth. At the same time, a fairly closed
canopy might prevent knotless parts of the stem from being devalued. Here,
positive and negative effects arise simultaneously, and thinnings are necessarily irrelevant whenever the positive equal or outweigh the negative effects.

The same applies to heterogeneous stands. Here, however, the potentially
differing growth rates have to be considered additionally (cf. Section 4.2.2).
Therefore, unequal growth might be compensated for by unequal impacts on
the remaining trees. Trees might be hold despite their low volume growth
rates in order to appreciate the positive effects on the unit timber revenues
of the remaining trees. Thus, positive unit revenue correlations equally restrict the relevant range of thinnings. In contrast to the homogeneous stand,
however, trees might be thinned even though they exert a net positive influence because of their own low growth rate.

4.4.1.2   Optimal Thinning Regime with Price Differentials

With price differentials, the optimal thinning regime is influenced. All factors
which potentially allow influencing the value of the next tree to be thinned
favor less intense or more frequent thinnings (cf. Sections 4.3.1 and 4.3.2).

Likewise, the optimal thinning intervals tend to be shortened over the age with degressively increasing unit revenues as the value growth is declining at a higher rate while the opposite applies to progressively decreasing unit revenues.

With reference to the thinning method, two trees with different value growth rates might be growing at equal volume rates but at different unit revenue rates. For instance, two equally vigorously growing trees might be shaped differently such that the unit revenue increase of badly shaped trees (e.g. wolf trees) might approach zero while the other increase is positive. Moreover, the impact of differently growing trees on the unit revenue of the remaining trees might diverge. The classic example is the poorly growing, overtopped tree which prevents knots from being formed on the stem of its dominant neighbor while the other, co-dominant neighbor severely competes with the dominant tree for resources. Here, incentives to cut the less vigorously growing tree previously arise.

With price differentiations, the determination of the optimal thinning method is then specified to lead to, cf. [4-47],

$$\left(\frac{p_{t_{k+1}}^{k+1}}{p^{k+1}} - \frac{p_{t_k}^{k}}{p^{k}}\right) + \left(\frac{q_{t_{k+1}}^{k+1}}{q^{k+1}} - \frac{q_{t_k}^{k}}{q^{k}}\right) > \frac{p_{t_k}^{k+1} q^{k+1}}{p^k q^k} + \frac{p^{k+1} q_{t_k}^{k+1}}{p^k q^k}$$

$$+ \sum_{j=k+2}^{n} \left[\left(\frac{p_{t_k}^{j} q^{j}}{p^k q^k} - \frac{p_{t_{k+1}}^{j} q^{j}}{p^{k+1} q^{k+1}}\right) + \left(\frac{p^{j} q_{t_k}^{j}}{p^k q^k} - \frac{p^{j} q_{t_{k+1}}^{j}}{p^{k+1} q^{k+1}}\right)\right] e^{r(t_k - t_j)}, \qquad [4\text{-}53]$$

evaluated at age $t_k{}^*$. Accordingly, volume and value growth rates might be distinguished. If a stand grows solitarily, the right hand side of [4-53] is zero. Without price differentials, the stand would then always be thinned from below since there are no incentives to hold poorly growing trees. Naturally, thinnings from below might here be employed only metaphorically since the Kraft (1884) classes are only relevant in competitive stands.

When the trees compete for resources, the right hand side of [4-53] might be positive or negative. Without price differentials, however, the net unit revenue rates as well as all prices might be cancelled out. In this case, only the development of timber growth remains as the determinant for the optimal thinning method. Following the timber growth theory (cf. Chapter 2), high Kraft (1884) classes exert the highest absolute impact in a competitive stand since the removal of their voluminous crowns release more resources than the removal of trees with small crowns. However, due to the disproportionately higher efficiency of dominant trees (cf. Section 4.3.4), even the impact rate of higher tree classes might be assumed to be of a greater magnitude since they transfer this efficiency to lower tree classes.

If it is thus assumed that the higher Kraft (1884) classes both grow and influence other trees with a higher rate, the right hand side of [4-53] is necessarily negative for thinnings from above if price differentials are absent. In this way, the question arises whether the higher impact rate of higher tree classes compensates for the loss of higher volume growth rates on the left hand side of [4-53]. Following the timber growth theory (cf. Chapter 2), the answer is negative since the contrary would imply that thinning from above leads to a higher timber volume compared to thinning from below. This case does not apply since the removal of bigger crowns is linked with lower resource utilization per unit area. Several experimental plots point in this direction (cf. Assmann 1970, p. 223 ff.; Nyland 2002, p. 231 ff.). Consequently, thinnings from above are irrelevant for investment situations without price differentials in this simplified setting.

If price differentials are present, nonzero unit revenue rates might offer incentives to thin forest stands from above. For solitarily growing stands, it is necessary that the difference in the unit revenue rates is inverse to and compensates for the difference in the volume growth rates. Thus, faster growing trees of low quality might be harvested prior to slower growing trees of high quality as long as the unit revenue difference outweighs the volume difference. In either way, the influence of the relative land rent has to be surpassed such that the whole stand is not clear-cut (cf. Section 4.2.2).

Differences in the unit revenue rates are dependent on the course of the unit revenue rates. If all trees face the same unit revenue function, i.e. $p^i(z) = p(z)$, the difference in the rates is zero and thus negligible even with price differentials. The same holds for different unit revenue functions with slopes proportional to the differences in the unit revenues i.e. $p^i(z) = \alpha^i p(z)$. In these cases, stands are thinned from below necessarily. Nonzero differences in the unit revenue rates are given for equal unit revenue changes and unequal unit revenues, for the reverse case as well as for disproportional changes. Therefore, high quality trees of lower classes do not only require an absolute advantage in unit revenue increases but the comparative advantage in relation to the already high timber price to induce thinnings from above.

Often, unit revenues are diameter-dependent which in turn are age- (and initial density-) dependent. For instance, unit revenues are frequently observed to increase in diminishing increment over rising diameters. Typically, diameter growth follows a sigmoid course over the age. The development of the unit revenues over the age is thus modified such that increasing unit revenue rates might be relevant at younger ages. Since less vigorously growing trees are faced with lower absolute diameter increments, the higher unit revenue rate per diameter unit of thinner trees has to exceed the higher absolute diameter increment of thicker trees in order to induce thinnings from above. Consequently, progressively increasing, regressively decreasing and linear unit revenues over rising diameters tend to favor thinnings from below (again in its metaphorical use). It thus pays to concentrate on the thickest trees as these promise the highest absolute and relative increase in timber value. If, on the other hand, unit revenues rise degressively, or decrease progressively respectively, over the diameters thinnings tend to be conducted from above (metaphorically) as the greatest opportunities for the improvement of the timber value lies potentially in the thinner trees.

When interdependencies between the trees enter the scene, the unambiguous determination of the optimal thinning method is complicated. Next to the timber volume growth and the unit revenue rates of the trees, the impact rates on both the timber volume and the unit revenues on all other trees have to be considered. Consequently, more vigorous growing trees with equal or higher unit revenue rates compared to less vigorously growing trees might be cut sooner if their negative impact on the remaining trees is great enough. Plausible situation might easily be constructed. Considering only two trees of deviating growth rates and diameters, the more vigorously growing tree might significantly restrict the timber growth of the less vigorously growing tree. With degressively increasing unit revenues over rising diameters, the impact rate of the more vigorously growing tree might be considerable higher than the one of the less vigorously growing tree since both the absolute increases in diameter and the relative increases in unit revenue per diameter unit are comparatively high.

Nevertheless, since the removal of higher tree classes exert more influence on the remaining trees than the removal of lower tree classes, degressively increasing, progressively decreasing and linear unit revenues tend to favor thinnings from above while progressively increasing and degressively decreasing unit revenues favor thinnings from below. Furthermore, positive interdependencies between trees might evolve when unit revenues are allowed to vary with the harvest ages. Typically, low tree classes might be kept in the stand in order to preserve the high quality of some dominant trees. In this case, the unit revenue impact rate on the right hand side of [4-53] is positive and thus appears as an additional marginal revenue. For thinnings from above, this tends to lower the right hand side of [4-53] thus tending to satisfy the inequality.

The previous considerations apply to continuous unit revenue functions. With discrete functions, i.e., with different prices for different timber sorts, unit revenue change rates are either zero or as high as the leap to the price of the next timber sort. In this way, it must be ascertained whether unit revenue jumps are relevant for the corresponding thinning or not. If, for instance, the potentially thinned tree and the remaining trees are far from a unit revenue leap, it might be profitable to concentrate more timber growth on the remaining trees if their product price is higher than the thinning product price. In either case, tough, the basic argument remains valid.

In summary, price differentials offer incentives to thin forest stands as another opportunity to increase the value of the remaining trees arises. In this sense, it might pay to concentrate timber growth on the trees with the highest value increment relative to the thinned timber value. However, trees considered for thinning might also yield unit revenue rates, which in turn lower the relevance of thinnings again. If the order of the trees according to the value growth rates deviates from the order of the trees according to the volume growth rates (tree classes), thinnings from above are conducted while, in the opposite case, the stand will be thinned from below. Degressively increasing, progressively decreasing and linear unit revenues over rising diameters tend to change the order of value growth rates compared to volume growth rates.

4.4.2   Harvest Cost

Any harvest of trees incurs costs to the forest owner. Naturally, these costs comprise fixed and variable costs. While the fixed costs are irrelevant for the isolated determination of the optimal harvest amount (Duerr 1993, p. 119), they are crucial for the determination of the optimal management regime due to the simultaneous equation system [3-13] - [3-15] as well as for the limits of profitable timber production. Variable costs, on the other hand, directly determine the optimal harvest ages as they depend on the order of harvests.

4.4.2.1   Variable Harvest Cost

Variable harvest or extraction costs accrue for every harvested volume of timber. The net unit timber revenue of a specific tree p^i for a harvested volume of timber is thus given by

$$p^i (z) = \rho^i(z) - h^i(z), \tag{4-54}$$

where ρ^i is the unit timber revenue, h^i are the variable harvest cost, and z is a vector comprising all relevant variables. Both the unit revenue and the variable harvest cost may be dependent on the harvest ages. Typically, the unit revenue varies with the timber structure as described in the preceding Section 4.4.1. For instance, thicker trees often yield higher timber prices per timber volume.

The variable harvest costs, on the other hand, vary in a specific way with the diameter of a tree. For equally long and formed stems, the number of stems

y which are required to sum a particular timber volume V is decreasing over the diameter of the stems d (all measured at the same location) since, cf. [2-4],

$$y = \frac{4V}{bld^2\pi},$$ [4-55]

where b is the form factor for the shape of the stem, and l is the length of the stem. For equal logging techniques, which imply equal costs for the harvest of one stem, the harvest costs are thus negatively correlated with the corresponding diameter according to a power law. As long as stands of the same age are compared, the tree height exerts no influence as it is assumed constant (cf. Section 2.1.3). For stands of different ages, however, the decline of the harvest cost is modified by the tree height. Since trees in dense stands are taller than younger and equally thick trees in less dense stands, all other things being equal, the increase of the harvest costs is attenuated, though not reversed due to the power functional relationship. The development of the adjustment factor, one the other hand, may attenuate or reinforce the decline.

Since the diameters of the trees in a stand are endogenously determined by the harvest ages (cf. Section 2.1.2), the variable harvest cost are dependent on these, i.e.,

$$h^i = h^i(t_1, \dots, t_i).$$ [4-56]

As postponed harvests are associated with additional diameter growth of the potentially thinned trees, while postponed thinnings reduce the diameter of the remaining trees (cf. Section 2.1.2), the changes in the harvest cost are given by

$$\frac{\partial h^i(t_1, \dots, t_i)}{\partial t_i} < 0$$ [4-57]

$$\frac{\partial h^i(t_1, \dots, t_i)}{\partial t_k} > 0 \quad with\ t_k < t_i.$$ [4-58]

In contrast to [4-57] and [4-58], the changes in the unit timber revenue due to changes in the timber structure are not unambiguous (cf. Section 4.4.1). If unit revenues are constant or increasing with larger diameters, the change in the net unit revenue p^i due to changes in the harvest ages points necessarily in the opposite direction of the changes in the harvest cost in [4-57] and [4-58], i.e., the net unit revenue a tree may obtain rises with its postponed harvest and decreases with postponed harvest of its neighboring trees. For decreasing unit revenues of thicker trees, the change depends on the relative magnitudes of unit revenue and harvest cost.

From this perspective it follows that thinnings always provide an argument for reducing the variable harvest costs. This relationship becomes relevant in the condition for the optimal thinning age [4-13]. Since the second term on the right hand side denotes the change in the timber value of all remaining trees, the likelihood to thin a tree increases when the variable harvest cost of future harvests can be reduced. In the same way, the relevant range for thinnings is influenced. Since the relevance of thinnings is determined with the help of the potential impact of thinnings on the remaining trees, cf. [4-20] for the homogenous and [4-25] for the heterogeneous stand, the relevance increases in the presence of variable harvest costs as the impact on the value of the remaining trees increases (cf. Paragraph 4.2).

The degressively decline of the variable harvest cost with increasing diameters is relevant in every stand. Without timber price differentials, net unit revenues thus increase over the age even in forest plantations for pulpwood where no structural timber aspects are relevant since the diameter growth follows a sigmoid course. As a consequence, the impact of the variable harvest costs is identical to the impact of digressively increasing unit revenues discussed in the preceding Section 4.4.1.2. Therefore, thinnings, but only comparatively late thinnings, become more relevant. Furthermore, thinnings tend to be less intense, conducted at shorter intervals and from above (cf. Section 4.4.1)

Without price differentials, trees of lower volume growth rates might grow in value at higher rates in the presence of variable harvest cost. As Appendix 1 (Paragraph 7.1) demonstrates for two trees with equal volume growth rates, the advantage of a higher absolute diameter increment of a thicker tree is overcompensated by the advantage of the steeper decline of the variable harvest costs of the thinner tree if [4-55] holds. Since, moreover, the higher net increment is set in relation to the lower timber volume and potential interdependencies between the trees favor the less vigorously growing tree more, the value growth rate of the thinner tree exceeds the one of the thicker tree necessarily. Thus, the thicker tree is harvested sooner. In this way, even lower volume growth rates may produce higher value growth rates if price differentials are absent. This relationship might be reinforced by degressively increasing or progressively decreasing unit revenues while it is attenuated in the reversed case.

In the presence of variable harvest cost, the net unit timber revenue might be negative. In these cases, thinnings are said to be precommercial. A thinning is denoted as precommercial if it holds at the age of the harvest t^p that

$$t^p := \{(t_1, \dots, t_i)|\rho^i - h^i < 0\}. \hspace{2cm} [4\text{-}59]$$

When thinnings are precommercial, the timber value of the tree turns negative since

$$p^i q^i = [\rho^i(t_1, \dots, t_i) - h^i(t_1, \dots, t_i)]q^i(t_1, \dots, t_i). \hspace{1cm} [4\text{-}60]$$

Hence, the costs of holding the tree value on the right hand side of [4-13] become the revenues of postponing the interest payments.

The corresponding value increment on the left hand side of [4-13] is then

$$\frac{\partial p^i q^i}{\partial t_i} = (\rho^i_{t_i} - h^i_{t_i})q^i + (\rho^i - h^i)q^i_{t_i}. \hspace{1.5cm} [4\text{-}61]$$

Here, two parts might be separated. During the time period ∂t_i, the changes in tree value due to a change in the net unit revenue and a change in the timber quantity. Since the change in the timber volume is necessarily positive (cf. Section 3.3.2), the second term of the value change in [4-61] is negative for precommercial thinnings as $\rho^i < h^i$. The change in the net unit revenue, on the other hand, depends on the relative magnitudes of the changes in the unit revenue and the harvest cost. For constant or increasing unit revenues, the change is necessarily positive as the change in the harvest cost is negative, cf. [4-57]. Only in the case of rapidly decreasing unit revenues, the net unit revenue change might be negative. In this situation, however, thinnings are irrelevant since timber production is unprofitable. Together with the positive timber volume, the first term of the value change in [4-61] is positive. As both changes are opposing each other, the total change in value is indeterminate in the case of precommercial thinnings.

The complete maximum condition for the optimal thinning age of the kth tree is then, cf. [4-13],

$$
\begin{aligned}
&\left(\rho^k_{t_k} - h^k_{t_k}\right)q^k + (\rho^k - h^k)q^k_{t_k} \\
&= r(\rho^i - h^i)q^i - \sum_{j=k+1}^{n} [(\rho^j_{t_k} - h^j_{t_k})q^j + (\rho^j - h^j)q^j_{t_k}]e^{r(t_k-t_j)}.
\end{aligned}
\qquad [4\text{-}62]
$$

Disregarding the impact on the remaining trees for the moment, condition [4-62] might be satisfied for negative magnitudes of both the value increment and the interest on the value as both are negative for precommercial thinnings. Figure 4.10 shows a fictitious example. The black solid curve illustrates the value increment. It is composed of both grey curves, which are the value change due to changes in the net unit revenue (solid) and due to changes in the timber quantity (dashed), cf. left hand side of [4-62]. Since the net unit revenue is zero at the beginning, the value change is negative first for a given precommercial phase. The same holds for the interest on the timber value, which is illustrated by the black dashed curve, cf. right hand side of [4-62]. Both the value change and the value interest cross each other in the negative quadrant. However, in the absence of mutual interdependencies between the trees, the second order sufficient condition [3-16] rules out this harvest age as it denotes a minimum due to the crossing of the value increment from below. If the tree growth is persistently poor, such that the tree does not leave the precommercial phase, its optimal thinning age is virtually

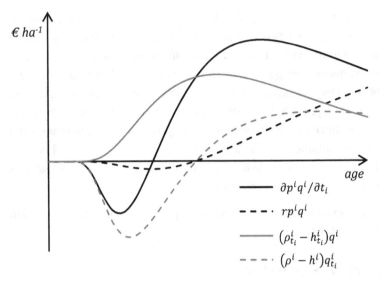

Figure 4.10 A fictitious example of precommercial thinning

infinite as its almost zero increment comes not in equality to the interest on its timber value. In this case, the costs of cutting the tree are postponed to the rotation age.

The presence of mutual interdependencies, though, might offer an argument for precommercial thinning. Since the unit timber revenue and the harvest costs as well as the timber quantity of the remaining trees might be positively influenced, cf. [4-62], it may pay to harvest trees although the cost might exceed the revenues. As long as the negative value increment exceeds the negative interest on the tree value, the impacts of postponing the thinning has to be positive in order to hold the tree. When the negative value increment is less than the negative interest on the tree value, it might nevertheless pay to thin a tree in order to generate the additional value of the remaining trees by increasing either or both the timber quantity and the unit revenue by decreasing the variable harvest cost. On the other hand, and despite the potentially negative impact on the neighboring trees, it might pay to hold undesired trees as their removal is too expensive compared to their impact. For instance, this might apply to unintentionally regenerated pioneer tree species.

Although precommercial thinning is an investment as capital has to be supplied in order to induce a future income, its occurrence is analogous to commercial thinnings in the model approach of this study. Like the regeneration cost at the beginning of a rotation period, precommercial thinnings regenerate the forest stand through a payout. On the one hand, the harvest of "negative value" by thinning precommercially will incur interest costs in the future which are higher the longer is the period between the thinning and the harvest of the remaining trees. This relationship offers incentives to postpone precommercial harvests to the age of commercial harvests. On the other hand, the earlier competitors are removed, the greater is the impact on the remaining trees. This might offer incentives to harvest trees just after the crossing of the threshold of competition. Depending on the opportunity to increase the value of the remaining trees, the first optimal thinning ages might be precommercial or commercial. However, due to the negatively inclined power functional relationship of the variable harvest cost and despite the negative net value, the unit revenue increase of precommercial trees is higher than the increase of commercial trees for equal diameter increases, which offers incentives to postpone thinnings in homogeneous stands and to thin from above in heterogeneous stands (cf. Section 4.4.1).

In summary, precommercial thinnings are irrelevant for both solitarily growing homogeneous and heterogeneous stands. In either case, trees will not be harvested as long as their timber value is negative. If the timber value will never reach a positive value, the optimal harvest age is infinite or coincides with the rotation age if timber production is profitable. With mutual interdependencies between the trees, it pays to thin trees precommercially if the impact on the remaining trees justifies the payments of the harvest and the accumulation interest costs.

4.4.2.2   Fixed Harvest Cost

In contrast to variable harvest cost, fixed harvest costs might occur with every thinning in an amount independent of the tree size. If the fixed harvest costs associated with the harvest of the ith tree are denoted by

$$H^i \begin{cases} = 0 & \textit{if } t_{i+1} = t_i \\ > 0 & \textit{if } t_{i+1} > t_i \end{cases},$$
[4-63]

they are only relevant if two adjacent harvest ages differ. The thinning model [3-6] can then be written as

$$LEV^H = (1 - e^{-rt_n})^{-1} \left[\sum_{i=1}^{n} (pq^i - H^i) e^{-rt_i} - C \right].$$
[4-64]

The associated optimal harvest age of the kth tree is determined by the maximization of [4-64] with respect to t_k, which yields after setting to zero and rearranging

$$pq_{t_k}^k + rH^k = rpq^k - \sum_{j=k+1}^{n} pq_{t_k}^j e^{-rt_j},$$
[4-65]

for $t_1, \dots, t_i = t_1^*, \dots, t_n^*$. Compared to the condition for the optimal thinning age without fixed harvest cost [4-13], the left hand side is extended by the interest on the fixed harvest cost associated with the harvest of the kth tree. Since these are necessarily non-negative, cf. [4-63], a postponement of the thinning might yield interest on the capital which must not be sacrificed for the harvest. The presence of fixed harvest costs thus increase the revenue side unilaterally thus increasing the optimal thinning age when viewed isolated from the equation system. The same holds for the optimal rotation age which is then determined according to, cf. [4-1],

$$pq_{t_n}^n + rH^n = rpq^n + rLEV^H.$$
[4-66]

Because the LEV decreases when fixed harvest cost are relevant while the revenue side increases, the optimal rotation age increases when viewed isolated from the equation system.

Since the fixed harvest cost of the ith tree are only relevant when it is not harvested simultaneously with the $i + 1$st tree, cf. [4-63], the optimal thinning intensity and the frequency, respectively, are influenced. With reference to the optimal thinning intensity in the homogeneous stand [4-36], two potentially successively harvested trees, k and $k + 1$, are harvested at the same age in the presence of fixed harvest cost if

$$-pq_{t_k}^{k+1} = rH^{k+1}. \qquad [4\text{-}67]$$

Accordingly, the two trees are harvested simultaneously if the impact of a postponement of the thinning on the next tree to be cut is less than or equal to the interest on the fixed harvest cost. In a solitarily growing and homogeneous stand, [4-67] is never satisfied with fixed harvest cost, and all trees are harvested at the rotation age.

The thinning is even intensified if more trees share the same relationship at the age t_k, i.e.,

$$-\sum_{i=k}^{k+j} pq_{t_i}^{i+1} = rH^{k+j}, \qquad [4\text{-}68]$$

where j is the number of trees with the corresponding investment characteristics. The same reasoning applies to the thinning frequency in the homogenous stand. If the impact on the remaining tree in [4-67], and in [4-68] respectively, is greater than the interest on the fixed harvest cost, the trees are cut successively.

In a heterogeneous stand, the condition for the optimal thinning intensity [4-37] becomes more complicated with fixed harvest cost due to the differing growth of the trees, i.e.,

$$\frac{pq_{t_{k+1}}^{k+1}}{pq^{k+1}} - \frac{pq_{t_k}^k}{pq^k} = \frac{pq_{t_k}^{k+1}}{pq^k} + \sum_{j=k+2}^{n} \left(\frac{pq_{t_k}^j}{pq^k} - \frac{pq_{t_{k+1}}^j}{pq^{k+1}} \right) e^{r(t_k - t_j)}$$
$$- \frac{rH^{k+1}}{pq^{k+1}}.$$

[4-69]

Hence, the impact of the differences in the relative influences on all remaining trees is reduced by the relative interest on the fixed harvest costs. Without mutual interdependencies, the right hand side of [4-69] is reduced to the term on the very right. As a result, two solitarily growing trees might differ in their growth rates but are cut simultaneously due to the relative interest on the fixed harvest cost. Or, from the perspective of the thinning frequency, differently and solitarily growing trees are only harvested at different ages if the deviation in the growth rates is high enough to offset the relative interest on the fixed harvest costs. Eventually, differences in the impact on the remaining trees might favor or handicap the thinning intensity, or frequency respectively. For instance, trees with similar growth rates but differing impacts on the remaining trees might nevertheless be harvest at the same age due to the presence of fixed harvest cost.

In summary, fixed harvest cost promote the thinning intensity while affecting the thinning frequency adversely since they offer incentives to balance losses of future additional timber volumes with gains from lower harvest costs. Especially in less competitive stands, the number of thinnings might be reduced severely by fixed harvest costs. Only considerable differences between the growth rates and/ or the impacts on the remaining trees offer incentives two cut trees successively when fixed harvest costs are comparatively high.

4.4.3   Regeneration Cost

Analogously to harvest costs, regeneration costs may appear in the form of variable and fixed costs, cf. [3-24]. In general, the impact of regeneration costs on the profitability of timber production was analyzed by Chang (1983). In his work, Chang was able to determine the changes in the optimal planting density and optimal rotation age due to changes in the investment parameters. However, while the changes due to variations of the fixed regeneration costs could be determined unambiguously, the change due to variations of the variable regeneration costs remain ambiguous and only determinable for some specified cases. In this way, further extensions of the model, as in Section 3.2.1, promise no greater insight as more, if not all, changes will become indeterminable. Therefore, the analysis here tries to combine aspects of the timber growth theory (cf. Chapter 2) and of the initial density model (cf. Section 3.2.1) in order to explore the implications for thinning and its relevance.

The initial density might be conceived as the planting density. In contrast to more complex regeneration methods like natural regeneration or sowing, problems like the impact of the storage and germination of the seeds as well as the distribution of the plants can then be omitted, which allows to focus on the relevant impact of the density. However, with each method, variable and fixed costs arise in some way which unifies the approach. The influence of the fixed regeneration costs is analogous to the impact of the fixed regeneration cost C employed in the analysis above, which therefore applies equally.

In a simultaneous equilibrium generated by the necessary condition [3-30], the optimal initial density m^* must satisfy, cf. [3-32],

$$n_m p q^n e^{-rt_n} = C_v - \sum_{i=1}^{n} p q_m^i e^{-rt_i}, \qquad\qquad [4\text{-}70]$$

with $m = m^*$. Accordingly, the additional timber value due to a change in the number of trees on the left hand side must equal the variable regeneration cost C_v corrected for the impact on the timber value of all already existing trees.

In view of the result in Sections 4.4.1 and 4.4.2, the effects of changes in the initial density might be specified. With reference to [4-54], [4-70] might be rewritten as

$$n_m(\rho^n - h^n)q^n e^{-rt_n} + \sum_{i=1}^{n} \rho_{t_i}^i q^i e^{-rt_i}$$

$$= C_v - \sum_{i=1}^{n} pq_m^i e^{-rt_i} + \sum_{i=1}^{n} h_{t_i}^i q^i e^{-rt_i}. \tag{4-71}$$

First, an additionally regenerated tree generates the opportunity to harvest an additional tree (first term on the left hand side). Naturally, the change in the number of trees due to a change in the initial density takes only integer values. Without mortality, $n_m = 1$. However, if the additional tree pushes the stand on the threshold of competition (cf. Figure 2.1), then $n_m = 0$ since one tree dies off in the course of competition. The second opportunity to increase the LEV by an additionally regenerated tree is given by potential unit revenue increases (second term on the left hand side). For instance, the additional tree might cause the timber quality to increase due to lower branch diameters or reduced areas of juvenile wood. Nevertheless, since unit revenues might also be affected by the dimension of the timber (cf. Section 4.4.1), the value of all trees might decrease due to thinner trees. In this case, the unit revenue effect is negatively inclined thus working as a marginal cost.

Obstacles to the regeneration of an additional tree are given on the right hand side of [4-71]. First, the variable regeneration cost C_v must be raised. Second, costs arise due to lower timber volumes of all trees within the range of competition as a result of smaller diameters (cf. Section 2.1.2). Third, and for the same reason, the variable harvest costs increase (cf. Section 4.4.2.1). Within the range of solitary growth, both terms on the very right side are zero. However, the management of solely solitarily growing trees is irrelevant (cf. Section 4.2.1).

Given positive variable regeneration cost, the right hand side of [4-71] is necessarily positive in all relevant cases. In order to ensure the balance, the left hand side must be positive equally. Assuming that unit revenues are constant over the age for the moment, i.e. in the presence of a unique timber price or unit revenue function, the opportunity of receiving gains by an additionally regenerated tree is reduced to the additional tree value. This additional value is zero in two cases: on the threshold of competition and for the equality of the unit revenue with the variable harvest cost. In either case, the marginal revenues of regenerating another tree cease while unilaterally imposing costs.

Typically, the stem diameter which realizes a timber price that just covers the variable harvest costs is referred to as the cost-covering diameter, or formally

$$d^{cc} := \{(t_1, \dots, t_n, m)|(\rho^i - h^i)q^i = 0\}. \tag{4-72}$$

This concept accounts for the dependence of the variable harvest costs, and to some extent of the unit revenue, on the stem diameter (cf. 4.4.2.1). Naturally, the harvest ages and the initial density must be chosen to generate diameters beyond the cost covering. If the first thinning is conducted before or just as the cost-covering diameter is reached, the additional regenerated tree will only raise cost since the marginal revenues on the left hand side of [4-71] are negative or zero.

Anyway, even if the harvest ages and the initial density are chosen such that the cost-covering diameter is exceeded, the revenues from a potential harvest of the additionally regenerated tree might not cover the additional regeneration costs. Consequently, a total variable cost-covering diameter might be defined as

$$d^{vc} := \{(t_1, \dots, t_n, m)|(\rho^i - h^i)q^i e^{-rt_i} - C_v = 0\}. \tag{4-73}$$

Any initial density may then not be chosen which generates stem diameters which fall short of the total variable cost-covering diameter at the age when mortality takes place. Additional trees may generate net revenues from thinnings but are unable to cover their accreted regeneration costs thus lowering the *LEV* necessarily. Fewer trees will further reduce the regeneration as well as the harvest costs while simultaneously increase the timber volume of all trees but will also reduce the opportunity to harvest more trees which in some investment situations might outweigh the cost.

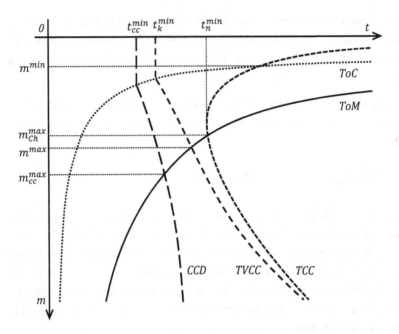

Figure 4.11 The relevant range of optimal initial densities

Eventually, in order to account for all costs, the prorated fixed regeneration cost might be added to define a total cost-covering diameter as

$$d^{tc} := \left\{ (t_1, \dots, t_n, m) \middle| (\rho^i - h^i)q^i e^{-rt_i} - C_v - \frac{C_f}{m} = 0 \right\}. \qquad [4\text{-}74]$$

This conception helps determining the boundaries of profitable timber production in general since any combinations of the harvest ages and the initial density which are unable to generate diameter above the total cost are irrelevant.

The preceding propositions are summarized in Figure 4.11. It is based on the results of the timber growth theory in Figure 2.1. Accordingly, combinations of the stand age and the initial density mark the thresholds of competition (ToC) and mortality (ToM). The long dashed line represents all combinations where the cost-covering diameter (CCD) is reached while the medium dashed line shows the total variable cost-covering diameter ($TVCCD$) combinations. Within the range of solitary growth, both lines are parallels to the initial density as the diameters remain unaffected. Beyond the threshold of competition, higher ages are necessary to guarantee equal diameters. Likewise, the distance between the curves of the CCD and the $TVCCD$ increases progressively as a result of the compounded interest on the variable regeneration cost. Since it takes a longer time to produce trees thick enough to cover their regeneration costs, this time elapsed exponentially increases the interest costs on the regeneration. At some point, higher initial densities might not be able to compensate for their regeneration costs such that the $TVCCD$ curve is parallel to the stand age.

The intersection point of the $TVCCD$ curve and the threshold of competition marks the maximal relevant initial density (m^{max}). Higher densities may produce trees which can be harvested with a net revenue starting from the crossing of the cost-covering diameter (m_{cc}^{max}), but which are unable to cover their regeneration costs. With higher interest rates, the relevant maximum moves to lower initial densities due to the compound interest. For m^{max}, thinnings are inevitable as otherwise trees are dying of which could have generated a net revenue. Due to the overall static approach, thinnings only serve to anticipate mortality before the $TVCCD$. After the crossing of the $TVCCD$, however, additional net revenues are generated which might compensate for the loss of the smaller diameters. In general, linear and progressively rising net unit revenues for thicker trees will make thinnings irrelevant as the planting density is reduced to allow solitary growth over the entire rotation age in order to produce the thickest trees possible within the production period. For degressively increasing net unit revenues and low variable harvest cost, the optimal initial density approaches m^{max} and thinnings become increasingly relevant. In this way, thinnings allow to produce additional value by regenerating more trees while simultaneously minimize the loss due to the thinner trees.

The lowest ages which allow to yield a repayment (t_{cc}^{min}) or a net variable revenue (t_k^{min}) are all solitarily growing combinations of the harvest ages and the initial density. In this way, t_k^{min} marks the earliest optimal thinning age within the initial density model approach. It follows that precommercial thinnings are irrelevant in this setting since the planting density is determined by anticipating thinnings. Because thinnings might be viewed as a way to reduce the initial density at higher ages, it cannot work as effectively as reductions of the initial density since the time elapsed until the harvest age and, therefore, the response options of the trees are limited. Additionally, reductions of the initial density reduce costs while precommercial thinnings increase costs.

To account for the prorated fixed costs, the narrowly dashed line in Figure 4.11 represents the total cost-covering diameter ($TCCD$) combinations of the harvest ages and the initial density. It takes the familiar form of the U-shaped average cost curve since the initial density, in contrast to the stand age, is a factor of production. In this particular representation, however, the relevant value axis is not shown as all curves are projections (cf. Section 2.1.5). Naturally, the (correct) average cost curve is U-shaped with respect to the initial densities as low initial densities must bear comparatively high fixed costs whereas high initial densities must bear high accumulated variable costs. The (tilted) U-shaped $TCCD$ curve in Figure 4.11, on the other hand, follows from compound interest. Low initial densities soon cross the cost and total variable cost-covering diameter but cannot cover the fixed costs, which are amplified by the interest costs, due to the low stem numbers. Very low initial densities might never produce timber at a profit. High initial densities, on the other hand, might cover the fixed cost quickly but are increasingly unable to cover the variable cost such that very high densities are equally unable to produce timber at a profit. In this way, the $TCCD$ curve marks the zero line of the LEV.

The lowest optimal initial density which might conceivably generate a maximal LEV in some investment situations (m^{min}) is determined by the intersection of the $TCCD$ and the ToC curve. Lower densities are either unprofitable in general or irrelevant due to increasing returns to scale (cf. Section 4.2.1). It follows that the relevant range of optimal initial densities is delimited by both $TVCCD$ and the $TCCD$ curve. More trees planted might be interpreted as wasted plants while fewer trees might be understood as wasted land. The potentially lowest optimal rotation age is determined by the tangent to the *TCCD* curve which is parallel to the initial density axis. Both higher and lower initial densities require more time to cover all costs.

In the planting density model proposed by Chang (1983), the relevant range of the optimal planting densities is given by the intersection points of the *TCCD* with *ToC* (m^{min}) and the *ToM* (m_{Ch}^{max}). Since all trees are cut at the same age, it is unprofitable to produce trees which cannot be harvested and which cannot promote the value of the remaining trees. In general, the Chang (1983) model is restricted to all combinations of the rotation age and initial density which lie between the *ToC* and *ToM*. Without any unit revenue differentiating effects, all planting densities except the lowest are relevant. In this case, it might pay to regenerate many trees since even the smallest might be sold at a profit.

The preceding analysis remains valid if net unit revenue changes, cf. second term on the left hand side of [4-71], are introduced which favor thicker trees or where the net unit revenue increase due to higher diameters equals or outweighs net unit revenue decreases from lower diameters. If the net unit revenue for higher diameters decreases, for instance due to the dominant impact of the timber quality, higher planting densities and precommercial thinnings cannot be precluded. In this way, if the value of higher timber quality is only assumed to be high enough, i.e., if the unit revenue change is progressively decreasing, all planting densities might be justified as long as the quality criteria are not over-fulfilled (cf. Coordes 2013). In these cases, thinnings are relevant as they allow to induce the positive unit value effects of higher diameters once the quality criteria are met satisfactorily. The classical example here is the two stage production of high quality sawlogs where high initial densities are followed by releasing selected trees in the second part of the rotation period. The optimal age for releasing some trees is basically determined by condition [4-52].

In summary, tree volumes, variable harvest costs, unit revenue premiums for thicker trees, variable planting costs and natural mortality all offer incentives not to plant more trees than are only just standing when the total variable cost-covering diameter is reached. Decreasing unit revenues over rising diameters, on the other hand, might induce to regenerate more trees. In either of these cases, thinnings basically appear as the anticipation of natural mortality.

4.4.4   Optimal Timber Volume

In the model approach of this work, the management of a forest stand is reduced to the individual harvest ages of each tree. For that reason, other stand characteristics are the consequences of the harvest ages. As one of these characteristics, the optimal timber volume, or often referred to as the optimal timber stocking, of a stand is an indirect consequence of the optimal harvest ages, and not the direct control variable as in other approaches (cf. Chapter 1).

For all cases in which thinnings are irrelevant (cf. Paragraph 4.2), the optimal timber volume is given by the untreated stand development. Typically, this natural stand development follows a sigmoid course over the age (cf. Section 2.1.3) with increasing increments at the beginning of the development followed by a period of decreasing increments. The optimal timber volume thus increases over the age until the final rotation harvest is conducted. This optimal rotation timber volume is then determined by the Faustmann-Pressler-Ohlin theorem [4-2]. In order to produce the maximum of standing timber, the planting density has to be chosen such that the beginning of natural mortality coincides with rotation age (cf. Sections 2.1.3 and 4.4.3). For the maximal growth performance, the planting density has virtually to be infinite and mortality has to be anticipated.

In forest stands in which thinnings are relevant, the development of the optimal timber volume is modified by the harvest of parts of the stand. Every thinning necessarily and by definition reduces the timber volume of a stand at the age of the thinning as trees are removed from the stand. Since this reduction occurs at one age, the accumulated timber volume function of the stand follows then a discontinuous course. As single trees are assumed to grow continuously without any discontinuous leaps, the timber volume of the stand is less than the timber volume of an untreated and otherwise identical stand for a more or less long time period after the thinning. Naturally, the optimal timber development remains unaffected until the first optimal thinning age.

In homogeneous stands and for constant unit timber revenues, however, thinnings necessarily reduce the timber volume compared to its identical, but untreated counterpart permanently. If it is possible to reduce the timber volume by a fixed quantity while maintaining the timber increments, the *LEV* increases since the same amount of timber can be produced with lower capital costs. In this case, it pays to remove every tree in order to reduce the capital costs. However, since no trees produce any increments, this situation does not arise. If the increments even increase due to the thinning, such that the timber volume will be as high as in the untreated stand at some age, the *LEV* can be even more increased as more is produced with less input. Here, it also pays to thin all trees instantaneously in order to gain the higher increments and reduce the capital costs. This, though, amounts to clear-cutting the stand which would be determined by condition [4-1] since future regeneration costs and timber revenues have to be considered.

With reference to the condition for the optimal thinning age [4-13], an uninfluenced timber volume of the remaining trees would be identical to a solitarily growing tree, i.e., the second term on the right hand side of [4-13] would be zero. If this situation would mark the optimal thinning age, the tree would be thinned when its value growth rate equals the rate of interest. In the homogenous stand, this would apply to all trees which in turn conflicts with condition [4-1] as the optimal thinning age would exceed the optimal rotation age. The same applies if the impact of a postponement of the thinnings on the remaining trees would be positive such that the rate of interest on the right hand side of [4-13] would be reduced. Only the reduction of the timber volume at the next optimal harvest age offers incentives to only cut one tree.

Figure 4.12 illustrates the argument. The solid curve shows the timber volume development over the age of the untreated stand. At t_k a thinning takes

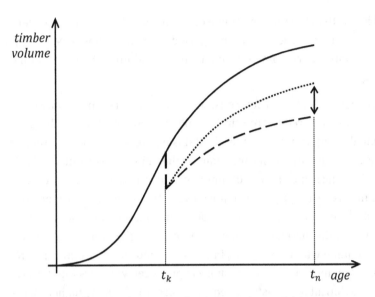

Figure 4.12 Effect of a thinning on the timber volume of the stand

place. It instantaneously reduces the timber volume of the stand by the timber volume of the thinned tree(s). If the trees are growing solitarily, the remaining trees continue to grow undisturbed (dashed curve) such that the timber volume at the age t_n is reduced proportionately to the thinning intensity at t_k compared to the untreated stand. With mutual interdependencies between the trees, the remaining trees grow more vigorously after the thinning due to the additional resources such that the timber volume rises compared to the solitarily growth (dotted curve). The additional timber volume at the age t_n induced by the thinning is then given by the difference between the competitive and the solitary growth. Basically, the thinned, competitive timber volume might exceed the untreated timber volume if mortality takes place between t_k and t_n (cf. Section 2.1.4). In a maximum of the *LEV*, however, this situation does not arise since it pays to thin the stand before as mortality can be utilized and capital cost can be reduced relative to the value increment.

In any case, optimal thinnings are not conducted in order to maximize the timber increments in any way. In contrast, the summed increment of the stand necessarily decreases on the average within a thinning interval. If the stand increment increases or remains unaffected, it is optimal to cut another tree since capital costs can be reduced without loss of revenues. Although the increments of the remaining trees increase in a competitive stand after a thinning, they are not maximized since this would imply solitary growth which is unprofitable in the homogeneous stand necessarily (cf. Section 4.2.1).

If thicker trees are more valuable such that the net unit timber revenue increases over rising diameters, for instance due to increasing unit revenues or to variable harvest costs, the optimal timber volume would be reduced even more compared to the untreated stand since the additional timber volume is realized by thicker remaining trees. As these are more valuable such that they incur higher opportunity costs, it pays to reduce the timber volume more strongly in order to generate higher value growth rates. For decreasing net unit revenues, on the other hand, all trees would be harvested at the same age in homogeneous stands. With correspondingly rising unit revenues, the optimal timber volume might be constant on the average or even decreasing. At some age, however, the optimal rotation age is reached as the number of trees is finite.

In heterogeneous stands, solitarily growing trees might be thinned (cf. Section 4.2.2). In this case, the optimal timber increments of the remaining trees remain unaffected such that course of the optimal timber volume is only reduced by the timber volume of the thinned tree. Within the range of competition, the optimal course of timber volume is reduced for constant or increasing net unit revenues. For decreasing unit revenues, the optimal timber volume might reach the untreated volume again or even surpass it in the thinning interval if natural mortality occurs in-between. In situations in which the projected timber increments would be constant or even increasing compared to the untreated stand, the optimal thinning intensity would not have been determined since the right hand side of condition [4-37] would exceed the left hand side.

In summary, optimal thinnings, whenever relevant, mine the timber volume of a stand over the age as they necessarily reduce both the timber volume and the timber increment. Although the individual increments of the remaining trees increase, the additional timber volume is not enough to compensate for the loss of timber volume at the thinning or the thinning intensity is less than optimal. Exceptions may be heterogeneous stands with decreasing unit timber revenues over rising diameters.

4.5   Comparative Static Analysis

With the help of a comparative static analysis, the influence of changes in the investment parameters on the optimal solution to [3-7] can be studied. Therefore, the total change in the optimal harvest ages due to an infinitesimal change in one parameter is observed. The relevant set of simultaneous equations, through which all parameter changes are induced, is given by equations [3-13] - [3-15]. At the point of a maximum $(t_1^*, \dots, t_n^*)$, they might be reduced to

$$\left. \frac{\partial LEV}{\partial t_1} \right|_{(t_1^*, \dots, t_n^*)} = -rq^1 + \sum_{j=1}^{n} q_{t_1}^j e^{r(t_1 - t_j)} = 0 \qquad [4\text{-}75]$$

$$\vdots \qquad\qquad \vdots$$

$$\left. \frac{\partial LEV}{\partial t_k} \right|_{(t_1^*, \dots, t_n^*)} = -rq^k + \sum_{j=k}^{n} q_{t_k}^j e^{r(t_k - t_j)} = 0 \qquad [4\text{-}76]$$

$$\vdots \qquad\qquad \vdots$$

$$\left. \frac{\partial LEV}{\partial t_n} \right|_{(t_1^*, \dots, t_n^*)} = pq_{t_n}^n - rpq^n - rLEV = 0. \qquad [4\text{-}77]$$

Throughout this section, assumption [3-47] is assumed to hold, i.e. the attention is restricted to competitive forest stands. Conditions [4-75] - [4-77] constitute a system of n equations with n unknowns. If the second order condition [3-16] holds at a point of a maximum $(t_1^*, \dots, t_n^*)$, the implicit function theorem can be employed since all functions possess continuously differentiable derivatives by definition (cf. Section 3.3.2), and since the Jacobi determinant, i.e.,

$$|J| \equiv \begin{vmatrix} \dfrac{\partial^2 LEV}{\partial t_1^2} & \cdots & \dfrac{\partial^2 LEV}{\partial t_1 \partial t_k} & \cdots & \dfrac{\partial^2 LEV}{\partial t_1 \partial t_n} \\ \vdots & \ddots & \vdots & \ddots & \vdots \\ \dfrac{\partial^2 LEV}{\partial t_k \partial t_1} & \cdots & \dfrac{\partial^2 LEV}{\partial t_k^2} & \cdots & \dfrac{\partial^2 LEV}{\partial t_k \partial t_n} \\ \vdots & \ddots & \vdots & \ddots & \vdots \\ \dfrac{\partial^2 LEV}{\partial t_n \partial t_1} & \cdots & \dfrac{\partial^2 LEV}{\partial t_n \partial t_k} & \cdots & \dfrac{\partial^2 LEV}{\partial t_n^2} \end{vmatrix} \qquad [4\text{-}78]$$

is nonzero by virtue of $|J| = |H|$ at $(t_1^*, \dots, t_n^*)$ in connection with [3-16]. Consequently, the optimal harvest ages can legitimately be interpreted as a function of the investment parameters, i.e.,

$$t_i^* = t_i^*(p, r, C). \qquad [4\text{-}79]$$

Since the equation system [4-75] - [4-77] can then be taken as an identity in some neighborhood of a maximum, the total differential of [4-75] - [4-77] with respect to $t_1^*, \dots, t_n^*, p, r, C$ can be written in matrix form as

$$\begin{bmatrix} \dfrac{\partial^2 LEV}{\partial t_1^2} & \cdots & \dfrac{\partial^2 LEV}{\partial t_1 \partial t_k} & \cdots & \dfrac{\partial^2 LEV}{\partial t_1 \partial t_n} \\ \vdots & \ddots & \vdots & \ddots & \vdots \\ \dfrac{\partial^2 LEV}{\partial t_k \partial t_1} & \cdots & \dfrac{\partial^2 LEV}{\partial t_k^2} & \cdots & \dfrac{\partial^2 LEV}{\partial t_k \partial t_n} \\ \vdots & \ddots & \vdots & \ddots & \vdots \\ \dfrac{\partial^2 LEV}{\partial t_n \partial t_1} & \cdots & \dfrac{\partial^2 LEV}{\partial t_n \partial t_k} & \cdots & \dfrac{\partial^2 LEV}{\partial t_n^2} \end{bmatrix} \begin{bmatrix} \left(\dfrac{\partial t_1}{\partial z}\right) \\ \vdots \\ \left(\dfrac{\partial t_k}{\partial z}\right) \\ \vdots \\ \left(\dfrac{\partial t_n}{\partial z}\right) \end{bmatrix} = \begin{bmatrix} -\dfrac{\partial^2 LEV}{\partial t_1 \partial p} \\ \vdots \\ -\dfrac{\partial^2 LEV}{\partial t_k \partial p} \\ \vdots \\ -\dfrac{\partial^2 LEV}{\partial t_n \partial p} \end{bmatrix}, \qquad [4\text{-}80]$$

where z may be any investment parameter p, r or C, while all t_i are evaluated at the maximum. Applying Cramer's rule, the comparative static derivatives can be expressed as

$$\left(\frac{\partial t_j^*}{\partial z}\right) = \frac{|J_j|}{|J|} \qquad j = 1,2,\dots,n.$$ [4-81]

In a maximum, the required second order derivatives, cf. [3-18] - [3-22], can be rewritten as

$$\left.\frac{\partial^2 LEV}{\partial t_k^2}\right|_{(t_1^*,\dots,t_n^*)} = \sum_{j=k}^{n} q_{t_k t_k}^j e^{r(t_k-t_j)} - r \sum_{j=k}^{n} q_{t_k}^j e^{r(t_k-t_j)}$$ [4-82]

$$\left.\frac{\partial^2 LEV}{\partial t_k^2}\right|_{(t_1^*,\dots,t_n^*)} = \sum_{j=k}^{n} q_{t_k t_k}^j e^{r(t_k-t_j)} - r \sum_{j=k}^{n} q_{t_k}^j e^{r(t_k-t_j)}$$ [4-83]

$$\left.\frac{\partial^2 LEV}{\partial t_k \partial t_1}\right|_{(t_1^*,\dots,t_n^*)} = -r q_{t_1}^k + \sum_{j=k}^{n} q_{t_k t_1}^j e^{r(t_k-t_j)}$$ [4-84]

$$\left.\frac{\partial^2 LEV}{\partial t_n \partial t_k}\right|_{(t_1^*,\dots,t_n^*)} = q_{t_n t_k}^n - r q_{t_k}^n.$$ [4-85]

The relevant derivatives with respect to the investment parameters are

$$\left.\frac{\partial^2 LEV}{\partial t_k \partial p}\right|_{(t_1^*,\dots,t_n^*)} = 0$$ [4-86]

$$\left.\frac{\partial^2 LEV}{\partial t_n \partial p}\right|_{(t_1^*,\dots,t_n^*)} = q_{t_n}^n - r q^n - r \frac{\partial LEV}{\partial p}$$ [4-87]

$$\left.\frac{\partial^2 LEV}{\partial t_k \partial C}\right|_{(t_1^*,\dots,t_n^*)} = 0$$ [4-88]

$$\left.\frac{\partial^2 LEV}{\partial t_n \partial C}\right|_{(t_1^*,\dots,t_n^*)} = r(1 - e^{-rt_n})^{-1}$$ [4-89]

$$\left.\frac{\partial^2 LEV}{\partial t_k \partial r}\right|_{(t_1^*,\dots,t_n^*)} = -q^k + \sum_{j=k+1}^{n} (t_k - t_j) q_{t_k}^j e^{r(t_k-t_j)}$$ [4-90]

$$\left.\frac{\partial^2 LEV}{\partial t_n \partial r}\right|_{(t_1^*,\dots,t_n^*)} = -p q^n - LEV - r \frac{\partial LEV}{\partial r}.$$ [4-91]

With these specifications, the impact of changes in the timber price (Section 4.5.1), in the regeneration cost (Section 4.5.2) and in the rate of interest (Section 4.5.3) can be indicated before a summary of the comparative derivatives is given (Section 4.5.4).

4.5.1   Price effect

Along with [4-81], the impact of a higher timber price on the optimal rotation age is given by

$$\left(\frac{\partial t_n{}^*}{\partial p}\right) = \frac{|J_n|}{|J|}.$$

[4-92]

Employing the Laplace expansion to the nth column, [4-92] can be expressed as

$$\left(\frac{\partial t_n{}^*}{\partial p}\right) = -|J|^{-1}\sum_{i=1}^{n}\frac{\partial^2 LEV}{\partial t_i \partial p}\left|C_{t_i p}^{J_n}\right|,$$

[4-93]

where $\left|C_{t_i t_j}^{J_n}\right|$ is the cofactor defined as $\left|C_{t_i t_j}^{J_n}\right| \equiv (-1)^{i+j}\left|M_{t_i t_j}^{J_n}\right|$, with $\left|M_{t_i t_j}^{J_n}\right|$ as the minor to the element of the ith row and jth column of the replaced Jacobian determinant $|J_n|$. Since the price impact on all optimal thinning ages is zero according to [4-86], [4-93] reduces to

$$\left(\frac{\partial t_n{}^*}{\partial p}\right) = -|J|^{-1}\frac{\partial^2 LEV}{\partial t_n \partial p}\left|C_{t_n p}^{J_n}\right|.$$

[4-94]

Given that $i = j$, the cofactor shares the same sign with its corresponding minor $\left|M_{t_n p}^{J_n}\right|$. Since the latter is equivalent to the $n-1$st order leading principal minor of the Hessian determinant, $|H_{n-1}|$, the sign of its value is given by virtue of the second order condition [3-16] as $(-1)^{n-1}$. The Jacobi determinant in the denominator of [4-94] is equal to the Hessian determinant $|H|$, and the sign of its value is therefore $(-1)^n$ by virtue of [3-16]. Together, the values of the minor and the Jacobi determinant have opposite signs thus turning the minus in front of [4-94] into a plus. The sign of the comparative derivative is then determined by the direction of the change in the optimal rotation age due to a rising timber price. According to [4-87], its change is negative because of $q_{t_n}^n - rq^n - r(1 - e^{-rt_n})^{-1} \sum_{i=1}^n q^i e^{-rt_i} = -p^{-1}r(1 - e^{-rt_n})^{-1}C < 0$. Hence, the optimal rotation age shortens when the timber price increases.

In the same way, the impact of a higher timber price on an optimal thinning age is given by

$$\left(\frac{\partial t_k^*}{\partial p}\right) = -|J|^{-1} \sum_{i=1}^n \frac{\partial^2 LEV}{\partial t_i \partial p} \left|C_{t_i p}^{Jk}\right| = -|J|^{-1} \frac{\partial^2 LEV}{\partial t_n \partial p} \left|C_{t_n p}^{Jk}\right|, \qquad [4\text{-}95]$$

when the t_k-replaced Jacobi determinant is expanded by the kth column. If k and n are both even or both odd, the sign before the minor $\left|M_{t_n p}^{Jk}\right|$ is positive while it is negative if k is odd and n is even, or *vice versa*. For an even n, the value of the Jacobi determinant is positive whereas it is negative if n is odd. Together, they thus turn into a positive value for even k, and into a negative value for odd k.

Due to the fact that the minor to the cofactor $\left|C_{t_n p}^{Jk}\right|$ is not symmetric, its evaluation is complex for large numbers of n. In order to assess the comparative static derivatives in more simplified approach, the problem is reduced to two tree classes which are characterized by an equal harvest age of all its trees. For instance, this would apply if the trees in the class grow homogeneously while hardly influencing each other negatively (cf. Paragraph 4.2), or if fixed harvest costs allow to cut trees in the stand at least twice (cf. Section 4.4.2.2). The maximization problem [3-7] is then reduced to [3-42] with the necessary condition given by the simultaneous equation system [3-43] and [3-44].

In this setting, the sign of the comparative static derivative of the thinning age t_k with respect to the timber price is determined by the minor to the cofactor $\left|C_{t_n p}^{J2}\right|$ which is given simply by [4-85]. If the change in the thinning induced additional increment of the rotation tree n due to a postponement of the thinning, i.e., $q_{t_n t_k}^n$, is positive, or if it is negative but less than the interest on the additional increment due to the thinning $(rq_{t_k}^n)$, the optimal thinning age will decrease as a consequence of an increasing timber price since k is odd. In the opposite case, the optimal thinning age will increase.

If three tree classes are distinguished analogously to [3-42], the sign of condition [4-84] becomes relevant. If it is positive along with [4-85], and if the thinning age to be considered is even, i.e., if the second thinning is considered, the corresponding minor to the cofactor $\left| C_{t_n p}^{J_k^2} \right|$ is positive, while it is negative if the corresponding thinning is odd. In combination with the sign in front of the minor, higher timber prices thus reduce the optimal thinning age. However, when [4-85] is negative, the change is ambiguous. For more tree classes, the unambiguity of the change depends, among other things, on the tree to be thinned.

Furthermore, the solitarily growing stand might be distinguished as another interesting special case. Here, trees grow independently of each other such that the cross-derivative of the harvest ages is zero, i.e. $\partial^2 LEV / \partial t_i \partial t_k = 0$. In this case, the Jacobi determinant [4-78] simplifies to a diagonal matrix, where all entries outside the principal diagonal are zero. Since the determinant of a diagonal matrix is the product of the entries on the principal diagonal, the sign of the cofactor in [4-94] remains unaffected. For the change in the optimal thinning age, on the other hand, the cofactor in [4-95] becomes necessarily zero as one row becomes zero. Therefore, the optimal thinning ages are independent of changes in the timber price in solitarily growing stands.

Yet another approach might be followed if two different timber prices for the two tree classes in [3-42] are distinguished. The LEV in [3-42] is then given by

$$
LEV = (1 - e^{-rt_n})^{-1} \left(\sum_{i=1}^{k} p^k q^i(t_k) e^{-rt_k} \right.
$$

$$
\left. + \sum_{j=k+1}^{n} p^n q^j(t_k, t_n) e^{-rt_n} - C \right), \qquad [4\text{-}96]
$$

where p^k is the timber price of the trees cut at t_k whereas p^n is the timber price for the trees cut at t_n with $p^k = p^n$ for $t_k = t_n$. Typically, trees harvested at earlier ages yield lower timber prices due to less valuable timber structures and higher harvest costs (cf. Sections 4.4.1 and 4.4.2).

In contrast to [4-76] and [4-86], the timber prices cannot be cancelled out since the necessary first order condition is given by the equation system

$$\frac{\partial LEV}{\partial t_k}\bigg|_{(t_k^*, t_n^*)} = \sum_{i=1}^{k} p^k q_{t_k}^i - r \sum_{i=1}^{k} p^k q^i + \sum_{j=k+1}^{n} p^n q_{t_k}^j e^{r(t_k - t_n)} = 0 \quad \text{[4-97]}$$

$$\frac{\partial LEV}{\partial t_n}\bigg|_{(t_k^*, t_n^*)} = \sum_{j=k+1}^{n} p^n q_{t_n}^j - r \sum_{j=k+1}^{n} p^n q^j - rLEV = 0. \quad \text{[4-98]}$$

Differentiating with respect to the timber prices yields

$$\frac{\partial^2 LEV}{\partial t_k \partial p^k}\bigg|_{(t_k^*, t_n^*)} = \sum_{i=1}^{k} q_{t_k}^i - r \sum_{i=1}^{k} q^i \quad \text{[4-99]}$$

$$\frac{\partial^2 LEV}{\partial t_k \partial p^n}\bigg|_{(t_k^*, t_n^*)} = \sum_{j=k+1}^{n} q_{t_k}^j \quad \text{[4-100]}$$

$$\frac{\partial^2 LEV}{\partial t_n \partial p^k}\bigg|_{(t_k^*, t_n^*)} = -r \frac{\partial LEV}{\partial p^k} \quad \text{[4-101]}$$

$$\frac{\partial^2 LEV}{\partial t_n \partial p^n}\bigg|_{(t_k^*, t_n^*)} = \sum_{j=k+1}^{n} q_{t_n}^j - r \sum_{j=k+1}^{n} q^j - r \frac{\partial LEV}{\partial p^n}. \quad \text{[4-102]}$$

Employing the Laplace expansion again, [4-92] can be expressed as the change in the rotation age due to a change in the thinning timber price as

$$\left(\frac{\partial t_n^*}{\partial p^k}\right) = -|J^2|^{-1} \sum_{i=k}^{n} \frac{\partial^2 LEV}{\partial t_i \partial p^k}\bigg|C_{t_i p^k}^{jn}\bigg|, \quad \text{[4-103]}$$

where $|J^2|$ refers to the two-tree-class Jabobi determinant, which is positive due to the second order condition, cf. [3-16]. Equivalently to the two-class-one-price case above, the direction of change depends on the minor to the cofactor $\left|C_{t_{np}}^{J_k^2}\right|$ which is analogous to [4-85]. If the thinnings induced increment change at the rotation age does not outweigh the interest on the additional timber volume, then the comparative static derivative in [4-103] is negative since [4-99] is necessarily positive in a maximum while [4-101] is negative. In this way, the optimal rotation age decreases due to a rise in the thinning timber price. If [4-85] is positive, however, the change remains ambiguous.

With respect to the change in the optimal thinning age, the comparative derivate is given by

$$\left(\frac{\partial t_k^*}{\partial p^k}\right) = -|J^2|^{-1} \sum_{i=k}^{n} \frac{\partial^2 LEV}{\partial t_i \partial p^k} \left|C_{t_i p^k}^{J_n^2}\right|. \qquad [4\text{-}104]$$

Again, the direction of change depends on [4-85]. If the latter is negative, the derivative is positive due to the opposing signs of the price derivatives [4-99] and [4-101]. Here, the thinning age increases due to an increasing thinning timber price. In the opposite case, the sign is indefinite.

The changes in the harvest ages by virtue of changes in the timber price at the rotation age, on the other hand, are less clear since the change in [4-102] depends on the relative magnitudes of the thinning class and the regeneration costs. If the discounted thinning value outweighs the regeneration costs, [4-102] is positive whereas it is negative only if the regeneration costs are predominant as is the case for the unique timber price. Therefore, with substantial thinning value in comparison to the employed regeneration capital, the isolated change in the optimal rotation age is just the opposite of the Faustmann model since the marginal revenues rise more than the marginal cost as only a part of the future timber revenues become more valuable.

The corresponding change in the optimal rotation age is

$$\left(\frac{\partial t_n^*}{\partial p^n}\right) = -|J^2|^{-1} \sum_{i=k}^{n} \frac{\partial^2 LEV}{\partial t_i \partial p^n} \left| C_{t_i p^n}^{J_n^2} \right|.$$ [4-105]

If [4-85] and [4-102] are negative, the change in [4-105] remains ambiguous as all partial changes are negative. The same holds in the opposite case. If [4-85] is positive and [4-102] negative, the optimal rotation age decreases with an increasing timber price p^n, and *vice versa*. Equivalently, the change in the optimal thinning age, i.e.,

$$\left(\frac{\partial t_k^*}{\partial p^n}\right) = -|J^2|^{-1} \sum_{i=k}^{n} \frac{\partial^2 LEV}{\partial t_i \partial p^n} \left| C_{t_i p^n}^{J_n^2} \right|$$ [4-106]

is ambiguous for both positive and negative [4-85] and [4-102]. For negative [4-85] and positive [4-102] as well as in the reverse case, the optimal thinning age decreases when the timber price at the rotation age increases.

Naturally, the changes in the different prices are relative to each other. Since an increase in the thinning timber price equally represents a relative decrease of the rotation timber price, and *vice versa*, the comparative static derivatives must show analogous relative changes. Interestingly, one approach is able to yield more specific results. An extension towards three tree classes is futile since either three prices must be distinguished or two prices spread over three classes. In either way, the comparative static derivatives remain ambiguous without further specifications.

For the solitarily growing stand, where $\partial^2 LEV / \partial t_i \partial t_k = 0$ and the Jacobi determinant becomes diagonally, the impacts of changes in the prices on the optimal thinning ages become irrelevant as the Laplace expansions of the thinning columns produce zero rows. Therefore, the optimal rotation age decreases with a rising thinning timber prices while it increases with rising rotation timber prices for $\partial^2 LEV / \partial t_n \partial p^n < 0$ and decreases in the reverse case. In the same way, the optimal thinning age rises with a rising thinning timber price while it necessarily decreases for rising rotation timber prices.

4.5.2   Regeneration Cost Effect

In the thinning model [3-6], the regeneration costs appear as a fixed payment at the beginning of each rotation cycle. With reference to the comparative static analysis, their influence is just the opposite of the general timber price effect since the corresponding partial derivatives just point in the other direction. The impact of the regeneration costs on the optimal rotation age is given by

$$\left(\frac{\partial t_n^*}{\partial C}\right) = -|J|^{-1}\frac{\partial^2 LEV}{\partial t_n \partial C}\left|C_{t_n C}^{J_n}\right| \qquad [4\text{-}107]$$

since the single optimal thinning ages are independent of the regeneration costs according to [4-88]. As in the case of the change in the optimal rotation age due to rising timber prices above, the direction of change is determined by the sign of the change in the optimal rotation age as a result of rising regeneration costs. Since this is positive with regards to [4-89], the optimal rotation age increases.

As in the case of the impact of the timber price, the unambiguity of the impact of rising regeneration cost on the optimal thinning ages is dependent on the number of trees. The optimal thinning ages of up to three tree classes increase when [4-84] and [4-85] are positive. In case the latter derivatives are negative, the change is ambiguous for three tree classes, and negative for two classes. For solitarily growing stands, the optimal thinning ages are independent of the regeneration costs (cf. also Section 4.5.1).

The comparative static changes due to changes in the regeneration costs can analogously be interpreted as the influence of fixed regeneration costs (cf. Section 4.4.3). Naturally, their impact is different form the impact of changes in the variable regeneration costs. As Chang (1983) has shown with the inclusion of the initial density as an endogenous variable, however, the comparative static analysis yields unambiguous results only with further specifications. In this way, further extensions of the model towards more harvest ages (cf. Sections 3.2.1 and 4.4.3) do not promise to gain deeper insight into the comparative static derivatives. Since in the initial density model (cf. Section 3.2.1) all first order conditions of the optimal thinning ages are independent of the timber price and the regeneration costs, the comparative static analysis is simplified. However, with the initial density, a second indirect channel is opened via which the thinning ages are influenced. The overall change cannot be determined except with several specifications.

4.5.3   Interest Rate Effect

The impact of a higher interest rate on the optimal rotation age is given by

$$\left(\frac{\partial t_n^{\ *}}{\partial r}\right) = -|J|^{-1} \begin{vmatrix} \dfrac{\partial^2 LEV}{\partial t_1^{\ 2}} & \cdots & \dfrac{\partial^2 LEV}{\partial t_1 \partial t_k} & \cdots & -\dfrac{\partial^2 LEV}{\partial t_1 r} \\ \vdots & \ddots & \vdots & \ddots & \vdots \\ \dfrac{\partial^2 LEV}{\partial t_k \partial t_1} & \cdots & \dfrac{\partial^2 LEV}{\partial t_k^{\ 2}} & \cdots & -\dfrac{\partial^2 LEV}{\partial t_k r} \\ \vdots & \ddots & \vdots & \ddots & \vdots \\ \dfrac{\partial^2 LEV}{\partial t_n \partial t_1} & \cdots & \dfrac{\partial^2 LEV}{\partial t_n \partial t_k} & \cdots & -\dfrac{\partial^2 LEV}{\partial t_n r} \end{vmatrix}. \qquad \text{[4-108]}$$

In contrast to changes in the timber price and the regeneration costs, a change in the interest rate is complex as it influences all optimal thinning age conditions, cf. [4-90], as well as the optimal rotation age condition, cf. [4-91]. As a consequence, any expansion of the replaced Jacobi determinant will include several non-symmetrical minors whose evaluation is problematic. Moreover, both changes in the conditions for the optimal thinning age, and the optimal rotation age respectively, are ambiguous without further specifications.

If the problem is reduced again to only two tree classes, the direction of change depends on the sign of the partial cross derivatives, [4-84] and [4-85], as well as on the sign of the changes in the maximum conditions in respond to the higher interest rate, i.e., [4-90] and [4-91]. If the former conditions are negative, the direction of change is ambiguous. If [4-84] and [4-85] are positive, the comparative static derivative shares the same sign with [4-90] and [4-91]. This analysis is also valid for three trees. Regarding the effect of a rising interest rate on the thinning ages, the analysis is similar due to the ambiguous sign of all derivatives with respect to the interest rate. However, the direction of change is already ambiguous for more than two tree classes even in view of all additional specifications. For the solitarily growing stand, the analysis is equivalent to the two and three class cases (cf. also Section 4.5.1).

As in other studies (e.g. Chang 1983; Li and Löfgren 2000; Halbritter and Deegen 2011), impacts of the interest rate are hard to assess. Since investments necessarily involve the balancing of payments occurring at different times, the different channels through which changes in the interest rate might travel multiple times with the number of payment dates. In this way, the unambiguity of the change in the rotation age due to a change in the interest rate in the Faustmann model (Amacher et al. 2009) is only a result of the stringent assumptions permitting the clear solution.

4.5.4 Summary

Table 4-1 summarizes the results of the comparative static analysis. For a typical forest stand of many trees, the comparative static analysis provides unambiguous results only for the change in the rotation age due to changes in the timber price and the regeneration costs as well as for solitarily growing stands. The influence of the interest rate, in contrast, as well as all shifts of the optimal thinning ages, remains ambiguous. For two or three trees, though, the direction of change is indicated when further specification of the timber growth are available. In particular, these are the changes in the timber increment at the optimal harvest age due to changes in previously conducted harvests. If, for instance, the increment increases and simultaneously outweighs the interest on the increment, rising timber prices and decreasing regeneration costs will shorten the optimal thinning age definitely for stands of two or three trees. For the complex effects of changing interest rates, though, additional specifications concerning their influence on the maximum conditions are necessary.

Table 4-1 Summary of the comparative static analysis

	$n \in \mathbb{N}$		$n = 3$		$n = 2$		
	$\dfrac{\partial^2 LEV}{\partial t_i \partial t_k} \lesseqgtr 0$	$\dfrac{\partial^2 LEV}{\partial t_i \partial t_k} = 0$	$\dfrac{\partial^2 LEV}{\partial t_i \partial t_k} > 0$	$\dfrac{\partial^2 LEV}{\partial t_i \partial t_k} < 0$	$\dfrac{\partial^2 LEV}{\partial t_i \partial t_k} > 0$	$\dfrac{\partial^2 LEV}{\partial t_i \partial t_k} < 0$	
$\left(\dfrac{\partial t_n^*}{\partial p}\right)$	< 0						
$\left(\dfrac{\partial t_k^*}{\partial p}\right)$	ambiguous	$= 0$	< 0	ambiguous	< 0	> 0	
$\left(\dfrac{\partial t_n^*}{\partial p^k}\right)$	ambiguous	< 0	ambiguous			< 0	
$\left(\dfrac{\partial t_k^*}{\partial p^k}\right)$	ambiguous	> 0	ambiguous			> 0	
$\left(\dfrac{\partial t_n^*}{\partial p^n}\right)$	ambiguous	> 0	ambiguous		ambiguous	> 0	$\dfrac{\partial^2 LEV}{\partial t_n \partial p^n} > 0$
		< 0			< 0	ambiguous	$\dfrac{\partial^2 LEV}{\partial t_n \partial p^n} < 0$
$\left(\dfrac{\partial t_k^*}{\partial p^n}\right)$	ambiguous	< 0	ambiguous		ambiguous	< 0	$\dfrac{\partial^2 LEV}{\partial t_n \partial p^n} > 0$
					< 0	ambiguous	$\dfrac{\partial^2 LEV}{\partial t_n \partial p^n} < 0$
$\left(\dfrac{\partial t_n^*}{\partial C}\right)$	> 0						
$\left(\dfrac{\partial t_k^*}{\partial C}\right)$	ambiguous	$= 0$	> 0	ambiguous	> 0	< 0	
$\left(\dfrac{\partial t_n^*}{\partial r}\right)$	ambiguous	> 0	> 0	ambiguous	> 0	ambiguous	$\dfrac{\partial^2 LEV}{\partial t_i \partial r} > 0$
		< 0	< 0		< 0		$\dfrac{\partial^2 LEV}{\partial t_i \partial r} < 0$
$\left(\dfrac{\partial t_k^*}{\partial r}\right)$	ambiguous	> 0	ambiguous		> 0	ambiguous	$\dfrac{\partial^2 LEV}{\partial t_i \partial r} > 0$
		< 0			< 0		$\dfrac{\partial^2 LEV}{\partial t_i \partial r} < 0$

5 Discussion and Conclusions

In the preceding analysis, the implications have been derived which follow from the simultaneous equation system developed in Chapter 3. In the present chapter, the analytical results are discussed in light of prior research and the questions expounded in Chapter 1. In order to retrace the incentives of forest owners thinning forest stands in the observable world, those situations have to be detected which are inconsistent with the model approach of this study. If these contradicting situations are observed, they are a strong indication of effective constraints deviating from those derived in this study. Through this process of "conjecture and refutation" (Popper 2002a), the true incentives might be approximated.

Accordingly, in the first paragraph, conclusions are drawn and discussed which follow directly from the model approach (Paragraph 5.1). Subsequently, the effects of relaxing the constraints of the analysis are discussed and extensions of the model indicated (Paragraph 5.2). In a third step, some applications of the model to interactive problems of conflict or cooperation are specified (Paragraph 5.3). Finally, the problem in this study is generalized to be applied to different land use concepts and entire forests (Paragraph 5.4).

5.1 Optimal Thinning

This paragraph seeks to discuss the implications which might be derived from the analysis of the preceding chapter. The focus lies on the optimal thinning regime, i.e., the discussion is based on the stringent assumptions of the "Faustmann laboratory" (Deegen et al. 2011, p. 363; cf. Section 3.3.1). First, the optimal cutting regime of a stand is viewed as the permanent balancing of harvests and regenerations (Section 5.1.1). Second, thinnings are reviewed as an attempt to control the density of a stand (Section 5.1.2). Third, the comparative static analysis is discussed against the background of the results of prior analyses (Section 5.1.3). In each section, conclusions are preceding the corresponding discussion.

5.1.1   Harvest and Regeneration

Thinning and reforestation are two possible ways to regenerate a forest stand

In the general form of [4-14], the determination of the optimal thinning age bears resemblance to the determination of the optimal Faustmann rotation age in [4-4], or the optimal rotation age in [4-3] respectively. In either case, the rates of value growth are balanced with the rate of interest. At the rotation age, however, the opportunity to regenerate the stand and to earn growth rates above the rate of interest, or more specifically, the interest on the land value in relation to the standing timber value, offers incentives to cut trees at higher growth rates. The land rent equals the interest on the *LEV* in a partial equilibrium (cf. Section 3.3.1). Since the *LEV* is the present value of bare forest land, which is defined as the infinite income stream generated by the production and sale of timber, cf. [3-6], the opportunity costs represent the alternative to regenerate the stand in order to start new rotation periods. This bears cost as it offers the opportunity to earn higher value increments in relation to the employed timber value (cf. Duerr 1960, p. 133).

In a similar way, a thinning might be interpreted as offering the opportunity to "regenerate" a forest stand by improving the growth rates of the remaining trees. This might be preferred to the alternative to clear-cut the stand and to earn increments from the reforestation of the stand when the additional increments of the remaining trees in the current rotation outweigh the revenues in future rotation periods, cf. [4-20] and [4-25]. Therefore, forest owners might regenerate their natural resource forest by reforestation or by thinnings. From an ecological perspective, both forms of regeneration are unquestionably different; however their economic consequences are similar.

Thinnings are investment and divestment at the same time

Optimal timber production is the balancing of regeneration and harvest. Each regenerated tree in an even-aged stand might be harvested to offer younger trees, or those of the same age, better growth conditions. At each harvest, future timber volume of the new or the remaining trees is compared with future timber volume of the harvested trees. Since the optimal rotation age (and, in a sense, the optimal thinning ages) simultaneously determines the optimal regeneration age, timber harvest and regeneration are constantly interlinked, and they are thus both investment and divestment at the same time. Next to harvests at the rotation age, thinnings thus offer the opportunity to control the investment in timber production (Smith et al. 1997, p. 91 f.).

In some situations, the investment character prevails while in other situations the divestment character is predominant (Deegen 2001, p. 10). The former is most evident for the regeneration at the beginning of the rotation period but also for precommercial thinnings where the cut trees might even remain unused in the stand. However, if the trees thinned might be utilized somehow, the divestment character is likewise relevant. At advanced ages, the divestment character, which is the transference of the timber capital to the next best alternative use, might eventually prevail. Then, thinning is more mining than regeneration. This mining character might also apply to solitary and heterogeneously growing stands (cf. Section 4.2.2). Nevertheless, even the final harvest at the rotation age is conducted in view of the opportunity to regenerate the stand. On the other hand, in situations in which the cost-covering diameter is comparatively low, for instance due to low wage rates, the divestment character might prevail already in fairly young stands. As in the calculation examples of Johann Heinrich von Thünen (1875, 2009), where already six year old trees yield net profits, planting 10,000 Scots pines can be most profitable when 8,000 of these trees can be sold as rods with profit after only a few years.

In forestry, these different investment characteristics are accounted for by the different terms for thinnings in differently stand ages. While the investment character dominates for precommercial thinnings (in German: "Pflegeeingriffe") where the focus lies on the growth of the remaining trees in young stands, commercial thinnings typically emphasize the divestment character in older stands (Smith et al. 1997, p. 113). In the German literature, commercial thinnings are often further subdivided into "Durchforstungen"

(thinnings in a narrow sense), where the divestment and investment character are more or less equally represented, and "Vornutzungen" (preceding
harvests), where the divestment character prevails.

When timber quality criteria are irrelevant, thinnings are primarily divestment

In the process of in- and divestment, the regeneration at the beginning of a
rotation period sets the scope of thinnings. Naturally, only regenerated trees
might eventually be thinned. In this way, forest owners plant trees in order
to anticipate future harvests by either thinning or clear-cutting as these determine the value of the forest stand, i.e. the value of the timber and the land,
either for exchange or for consumption or reinvestment, the latter being
equivalent under the Fisherian separation theorem (cf. Paragraph 3.3). For
this, gains from additional timber of each regenerated tree must be balanced
with losses from lower timber volumes of each single tree. Moreover, each of
these timber volumes might be differently valuable. Forest owners regenerating their forest stands with high initial densities thus either speculate on
early profits from thinnings or clear-cutting (as von Thünen in the example
above) or on high prices for high quality timber. If rotation ages are comparatively long, high initial densities imply low cost-covering diameters when
timber quality is not decisive (cf. Section 4.4.3). Without the opportunity to
thin, these stands could only be regenerated with low initial densities.

In general, comparatively low initial densities suggest thinnings to be assumed less relevant. Here, the often adduced price premiums for thicker
trees render thinnings irrelevant since they are fully exploited by the low
initial density. Only increasingly price-independent timber structures favor
then higher initial densities as more timber of equal value can be harvested.
Therefore, the combination of low cost-covering diameters and relatively
price-insensitive timber volumes favor high initial densities even in regions
with comparatively long rotation periods. For instance, low wage rates and
high prices for thin timber structures, such as firewood, might thus lead to
increasing initial densities.

If initial densities are high enough to induce thinnings without substantial influence of timber quality aspects, trees are thinned as they allow for transference of some capital to its next best alternative use while simultaneously reducing the value increment not proportionately but to a lesser extent, thus boosting the rate of return (cf. Section 4.4.4). In this way, not merely the highest value increment is relevant for thinnings to be conducted but its value increment set in relation to its employed value. If thicker dominant trees yield higher increments (cf. Nyland 2002, p. 414), it might be not enough to yield higher growth rates as the diameter must increase by the factor $\sqrt{a+1}$ (cf. Appendix 1, Paragraph 7.1) compared to its already accumulated diameter.

The rising capital costs per tree for employing the land prevent stands from being exploited

The more intense the competition for resources, i.e., the higher the potential impact of the harvest of a tree on the timber volume of the remaining trees, the more likely thinnings are (most) profitable as less trees are cut at the rotation age, cf. [4-20] and [4-25], respectively. Conversely, with low levels or without any competition, most or all trees will be cut at the rotation age. Basically, the intensity of competition is regulated by the initial density, i.e., the number of trees at the beginning of the rotation period. High densities, such as natural regeneration might produce, will increase the portion of thinned trees in contrast to comparatively low planting densities as the potential to influence remaining trees increases. Furthermore, the intensity of competition is dependent on the tree species. The effect of a harvest of a shade-tolerant tree on its conspecifics might be greater than in a stand of shade-intolerant trees at comparatively high age classes due to asymmetric competition pressures while the reverse might hold for lower age classes where competition in shade-intolerant stands is severe. If this applies, more trees are thinned in stands of shade-tolerant trees at later ages and earlier in shade-intolerant stands, all other things being equal. Or, the competition intensity might be site specific. Forest stands on sites of low soil quality or

growing in a disadvantageous climate are widely spaced as some scarce resources restrict the availability of those which are more available for the
trees. In these cases, thinnings might hardly influence the remaining trees.
Thus, all trees will be cut at the rotation age provided they are growing more
or less homogenously.

Conversely, it might be seldom advantageous to hold only few or one tree
until the final rotation age. In this case, according to [4-20] or [4-25], the
whole land rent is concentrated on one tree thus boosting the costs of holding the land, i.e., the costs of starting a new rotation with higher growth rates,
which in turn shortens the rotation age thus bringing it closer to the next
thinning age. Only if a stand is small and/ or the trees are huge, it may be
optimal to hold only one tree. Therefore, if the dominant trees are removed
due to their low value growth rates (selection thinning), and if the remaining
co-dominant and overtopped trees leave a somehow unsatisfactory impression due to large openings, the optimal rotation age might have been exceeded since the remaining trees are unable to justify the high relative land
rent. When these situations of degraded forests (Hyde 2012, p. 22 ff.) are
thus observed, they are a strong indication that forest users put little value
on the land as a source of future income (for instance due to unsecure property rights or legal prohibitions to appropriate future timber products).
Here, the investment opportunities are restrained such that thinning remains as the only form of harvest since regeneration of any kind offers no
incentives.

Due to the interdependence of the optimal harvest ages, the determination of the optimal harvest regime is a feedbacked two-stage process

The influence of the relative land rent emphasizes the interdependencies between the optimal harvest ages. The determination of the optimal harvest
ages is a simultaneous process. A harvest age is only optimal given that all
other harvest ages are optimal. The isolated optimization of harvest ages, on
the other hand, might lead to severe deviations from the global maximum.
For instance, if several trees are thinned due to their low value growth rates,

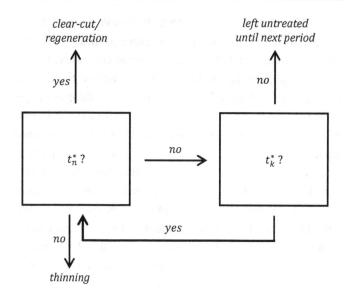

Figure 5.1 A two-stage approach towards the optimal cutting regime

the remaining stand might be unsatisfactorily sparse. Again, this impression of sparseness might be explained with the rising costs in the form of the relative land rent. The latter is relative to standing timber value, which might be severely reduced by thinnings thus boosting the relative land rent. These high costs might then not justify the prolongation of the current investment, which amounts to clear-cutting with subsequent regeneration. In this sense, the land costs per tree are too high due to fallow land within the stand. The simultaneous determination even holds without interdependent tree growth. If the latter enters the consideration, the other marginal determinants are also influenced through feedback processes between the maximum conditions.

Due to the two different kinds of harvest ages, the determination of the optimal cutting regime in a forest stand is basically a two-stage operation (cf. Figure 5.1). First, the optimal rotation age is assessed (t_n^* in Figure 5.1). If the transference of the timber value to the next-best alternative use and the re-generation with younger, faster growing trees is more profitable than the value increment of the standing timber, the stand is clear-cut and reforested (see Figure 5.1). If the optimal rotation age is not reached yet, the optimal thinning age is examined (t_k^* in Figure 5.1). If the optimal thinning age is yet (or never) to come, the stand is left untreated (see Figure 5.1). If, then again, the optimal thinning age of some trees is reached, it must be ascertained whether the removal of these trees does not reintroduce the optimality of the rotation age by virtue of the simultaneous determination process (see feedback arrow in Figure 5.1). If not, the corresponding trees are thinned (while ensuring that not more trees might be optimally thinned at the same time due to the simultaneous determination). And if so, the stand is clear-cut (see Figure 5.1). The approach might be even more generalized if t_n^* is sub-stituted for t_{k+1}^*, which emphasizes the dependence between all harvest ages.

Without the awareness of the interdependence of the harvest ages (espe-cially the feedback process in Figure 5.1) and thus of the two-stage approach, the most profitable cutting regime might be missed completely. This applies especially to the pragmatic approach of free tree-marking. In assessing the influence of different thinning variants on a group of trees within a forest stand, the overall impact of each group on the whole stand might easily be neglected. Both more or less intense thinnings as well as the final clear-cut might be evaluated differently after the thinnings are conducted according to the tree-marking. In this way, the interdependencies between the tree groups are disregarded.

The most profitable tree is a relative concept

When thinnings are profitable such that the feedback process in Figure 5.1 does not apply, the question arises which trees are to be removed. Basically, the answer to the optimal thinning ages determines the optimal thinning method (cf. Section 4.3.4). In contrast to exploiting the standing timber value, thinnings from above are only conducted when it is profitable to hold on to

both fast growing and slow growing trees while thinnings from below assume the slow growing trees to be comparatively unprofitable. The relationship between these two thinning methods emphasizes the relativity of fast
and slow growing trees. Dominant and co-dominant trees yield high returns
by virtue of their high growth rates, but they are likewise expensive in terms
of their high influence on the remaining trees. Dominated trees, on the other
hand, yield only little returns, but are comparatively cheap as they hardly
influence the value of their neighboring trees. The same holds for trees of
high and low quality: high value increments due to high quality might be relatively small in proportion to the high value employed for the increment
while small increases in value might promise satisfactorily value growth
rates if the tree value is low. Therefore, if the high marginal revenues of dominant trees relative to their high marginal costs in the form of high impact
rates are lower than the low marginal revenues of dominated trees relative
to their low marginal cost, thinnings from above are more profitable, i.e., it
pays to cut more vigorously growing prior to less vigorously growing trees.

Equal investment situations might lead to different thinning methods

The relativity of the most profitable trees challenges the unique applicability
of different thinning methods to specific classes of investment situations. In
contrast, similar price constellations might favor different thinning methods.
For instance, in expectation of rising premiums for thicker trees, the objective to concentrate increments on selected trees might favor heavy thinnings
from below as the individual growth of the already thickest trees is then maximized. However, if less vigorously growing trees exist which grow more or
less independently of the selected trees and have value growth rates above
the rate of interest or haven't crossed the cost-covering diameter yet, the optimal thinning might be conducted from above since the less vigorously
growing trees remain as they do not interfere with the other trees. Conversely, thinnings might also be conducted from above when the net unit revenues rise degressively over the diameter, such as in the case of the variable
harvest costs (cf. Section 4.4.2.1). In this situation, the less vigorously growing trees remain because they promise the greatest opportunity to increase

the total value the most. In either of these cases, small trees remain in the stand, however for entirely different reasons. In yet other situations, small trees might remain due to their positive impact on value of the remaining trees.

Optimal thinnings do not necessarily further increase the growth rates of the best growing trees

Thinnings from above are conducted whenever the order according to the value growth rates differs from the order according to the KRAFT (1884) classes. In this way, rising heterogeneity concerning the quality growth rates (cf. Section 4.4.1) will tend to favor high thinnings as in the example of wolf trees. Thinnings from below, on the other hand, demand the profitability of producing comparatively thick trees. If small diameters are highly valued, for instance as firewood or as an agricultural building material, it surely pays to harvest thin trees. In this situation, however, it might also pay to clear-cut the comparatively young stand entirely as in the case of coppice forests. Thinnings form below, on the other hand, are only profitable when the simultaneous production of thicker trees is also viable. The thin timber crops are then a joint product for which small additional diameter increments of the remaining thicker trees, but not the entire thick tree, are sacrificed in order to allow unthinned trees to develop. Since the unit volume relationship underlying the variable harvest costs (cf. Section 4.4.2.1) applies to every stand, thinnings from below are in no way necessarily connected to mass production of timber.

Therefore, optimal thinnings are not conducted in order to further increase the already high value growth rates of some trees necessarily. Optimal thinnings are conducted to gain the highest increase in the value growth rates of the whole investment, i.e., the current forest stand including all future stand states. Whether the growth rate increase is realized with hitherto poorly growing trees or with the best growing trees is not determinable *ex ante*. Thinning concepts favoring the (assumed) best growing trees therefore imply that the promotion of their growth rates yields higher increases than the promotion of both the less well growing trees and future, not yet regenerated trees. These high demands on the absolute value increments (due to the usually high value input of the best growing trees, cf. Appendix 1, Paragraph 7.1) might not be met by many trees in the stand.

With rising heterogeneity concerning the value growth rates, forest stands are thinned irregularly

The allocation of trees to different products, for instance to pulpwood and sawlog production (Smith et al. 1997, p. 117), depends on the availability of promising trees. In homogeneous stands, either sawlog production with (possibly severe) thinnings or pulpwood production without or with only slight thinnings is more profitable since all trees are growing equally and each part the optimal path must be optimal according to the principle of optimality (Bellman 1957). The uniform treatment of homogeneous stands is independent of the possibility that the stand grows heterogeneously afterwards, and, likewise, it is independent of potential heterogeneity in the prices for different timber structures. In heterogeneous stands, by contrast, only few trees might be profitable for sawlog production while most of the other trees are more profitable when producing pulpwood. In this process, optimal production of a single tree is always the best given what every other tree is producing by virtue of the simultaneous equation system, cf. [4-75] - [4-77]. All pulpwood trees might be converted to sawlog trees by selective promotion, but it might be not worth the cost of sacrificing other pulpwood trees.

Nevertheless, although heterogeneity regarding tree growth is necessary for unequal treatments of stands, it is not sufficient since homogeneous distribution of differently growing trees might entail a uniform treatment. This applies when different homogeneous stands are mixed regularly. With spatial heterogeneity due to substantial genetic or site variability, on the other hand, combinations of different timber products are likely to be optimal if the marginal rate of substitution between the products is not equal for every point in the stand, i.e., if the relative differences are not equal between all

trees. However, the concept of the forest stand is related to more or less homogeneous management conditions such that two or more stands might be
differentiated in some situations. In this way, production is specialized (cf.
Vincent and Binkley 1993), however in sub-stands.

5.1.2   Thinning and Density

**Thinnings are qualitatively equivalent to and quantitatively less pro-
nounced than reductions of the initial density**

Basically, thinnings are harvests of trees prior to the rotation age. In this way,
thinnings offer the opportunity to gain access to the control of density in a
forest stand. However density is defined or measured (cf. Zeide 2005; see
also Chapter 2), thinnings influence all density criteria independently of the
age of the stand, although conceivably to varying degrees. Even the notion of
"thinning" implies reductions of density. (Interestingly, the corresponding
German term "Durchforstung" is used in everyday language in the way of
"searching for something specific".) In this way, thinnings are sometimes referred to as the control of the relative density in a forest stand (cf. e.g. Nyland
2002, p. 402). If, on the other hand, timber growth is denoted density-dependent as the change in the timber volume is dependent on the current
stock size (cf. Conrad and Clark 1987, p. 62), thinnings are restricted to the
stands where mutual interdependencies between trees arise. As Section
4.2.2 has shown, however, thinnings might be conducted in heterogeneous
stands even with solitarily growing trees. Nevertheless, density might
equally be defined in terms of competitive pressure with rising density due
to rising impact rates.

While all density criteria might be differently influenced by thinnings, they
all yield equivalent results if applied to the initial density (cf. Chapter 2). Often, thinnings are viewed quite separately from the initial density. However,
both are unquestionably interlinked systematically. Without (enough) regenerated trees, thinnings become irrelevant. Furthermore, the frequency
and intensity of thinnings are predetermined as, at some age, no trees might
be available anymore. In principle, thinnings are qualitatively equivalent to

reductions of the initial density while postponements of thinnings work as increases of the initial density. Therefore, both entail equal consequences for the timber volume of a forest stand although the effect of thinnings is quantitatively less pronounced (cf. Section 2.1.4). In this way, the path of density is initiated by the initial density and adjusted by thinnings.

Nevertheless, the stand density is the consequence of the regeneration and of subsequent harvests, not the reverse. Attempts to optimize the density thus often yield ambiguous results as the trees to be removed cannot be derived. Besides, the response of the remaining trees might be totally different depending on the timber structure. For instance, the same timber volume might be comprised of a quarter of the trees of the same age however with twice the diameter or any combination in-between. The further development of these stands is surely unalike as wood does not grow wood. The composition of a given timber stand volume, though, is the result of the initial density and, potentially, of earlier conducted harvests. Only if all initial densities and subsequent treatments are assumed to be similar, comparisons can be meaningful. In the presence of price differentials and variable harvest cost, on the other hand, the timber structure as originated by the initial density becomes increasingly relevant since different volume growth rates might be accompanied by even more deviating net unit revenue growth rates. While density control might thus be suitable for analyses of aggregated timber production in forests (Borchert 2002, p. 66), it might be less appropriate for stand-based prescriptions.

Thinnings become potentially relevant through negative interdependencies concerning the tree values and heterogeneity concerning the rate of value growth

In the presented model [3-6], mutual interdependencies between trees are captured in the timber volume function of a tree [3-5]. Still, these interdependencies appear in two different ways. [3-5] might be separated to lead to

$$q^i = \varphi^i(t_i) + \sigma^i(t_1, \dots, t_i).$$

[5-1]

Here, φ denotes the thinning-unaffected timber volume of a tree while σ represents the additional timber volume due to earlier harvested competitors. Without previous thinnings, $\sigma = 0$. Alternatively, φ might denote the solitary growth of a tree which is reduced by σ representing the negative influence of the presence of competitive trees. Since both approaches are qualitatively equivalent, the first interpretation is followed in the analysis.

[5-1] distinguishes between the given and the additional timber volume produced. In the thinning model [3-6], the forest owner is able to influence the density via thinnings, which become potentially relevant through the additional timber volume of the remaining trees σ (cf. Paragraph 4.2). The existence and degree of this potential, though, is given implicitly by the exogenously determined timber volume φ. Next to previously conducted harvests, φ is basically determined by the initial density. Hence, it prescribes the development of a tree within the exogenously given boundary conditions. These might or might not comprise density-dependent aspects. With competitors effectively restricting the growing space, φ is simply of smaller magnitude than with less or without competitors. In the same way, the timber volume function of the Faustmann model [3-3] does not necessarily disregard density-dependent relationships as these might be built in, for instance by virtue of thinner trees compared to the same stand with less trees. This reasoning provides an argument against the sometimes applied contraposition of the Faustmann model to the fishery model (for the latter cf. e.g. Neher 1990).

In the light of this contemplation, the Faustmann model might be interpreted as a model for homogeneous and solitarily growing forest stands. In this setting, neither harvests nor regenerations prior to the rotation age might possibly increase the income from forest stands in an efficient market. Hence, all trees are necessarily cut at the same age. In view of positive land rents, even some degrees of heterogeneity lead to a uniform harvest age, cf. [4-25], which, in addition, might even be extended by fixed harvest costs, cf. [4-69].

With rising positive interdependencies between the values of the trees in a stand, the optimal harvest ages converge

Although the analysis in this study concentrates on the negative effects of joint timber growth of even-aged trees in a stand, i.e., $q^i_{t_h} < 0$ in [3-47], or $\sigma^i > 0$ in [5-1] respectively, the model covers the whole range of timber volume interdependencies between trees. When the narrow view of the solely competitive stand is abandoned, trees might improve each other's timber volume in some situations. Next to the negative influence of competitive pressure on the timber volume of a tree, situations with mutually reinforcing timber volumes might thus be analyzed. From the set of all interdependencies between trees, those are relevant for the forest owner which affect the tree value. For instance, this might also refer to the impact of protection against biotic and abiotic damages such as pests or storms.

All factors increasing the value of the remaining trees, such as lower harvest cost, higher timber prices or more timer volume, increase the likelihood of thinnings to be conducted as $p\sigma^i$ increases. Conversely, factors decreasing the value of the remaining trees, such as lower timber quality or less timber volume (in cases of mutually reinforcing timber growth), thinnings are less likely to be conducted. In the presence of positive effects on the remaining trees, or when the positive exceed the negative effects, the presented model will be reduced again to the Faustmann model as it always pays to prolong the thinning in order to avoid the negative impact of an early harvest. For instance, this situation might be relevant in comparatively dense stands where the closed canopy might prevent lower stem portions from being devalued by epicormic branching caused by the additional light shed by thinning gaps which outweighs the negative effects of lower tree volumes and higher harvest costs. In this way, the Faustmann model might equally be interpreted as assuming the negative influences to be low enough or not even existent or to be exceeded by the positive effects since in these instances all trees share the same optimal harvest age.

The reasons to thin a stand more intensively are given by equal and independent or unequal and interdependent value growth

If the opportunity to harvest timber is restricted to the possibility of harvesting trees, thinnings naturally reduce the density instantaneously by the proportion of the thinned trees. These removals of single trees causes the aggregated growth of the stand to follow a discrete course which is at best piecewise differentiable and thus not open to standard maximization approaches. If stand density is directly optimized, only increment equivalents may be removed of the stand since the aggregated timber increment of the stand is realized in different shares on each tree. However, the sharp reduction of the density, and thus of the timber volume respectively, might be even intensified if more trees are removed at the same age. When two trees share the same optimal thinning age, the thinning is conducted more intensively and [4-36], or [4-37] respectively, holds.

However, under which conditions is $q_{t_k}^{k+1}$ in [4-36] zero? $q_{t_k}^{k+1}$ is the impact of a postponement of the harvest of a tree on the next tree to be cut or the foregone additional increment of that tree. As a start, it is zero for solitarily growing trees. Without competition, trees in homogenous forest stands will all be cut at the same age, i.e., at the Faustmann rotation age according to [4-2]. In a competitive forest stand, by contrast, $q_{t_k}^{k+1}$ might be negative, cf. [3-47]. However, in competitive and homogeneous stands, the harvest of one tree will potentially affect the timber volume of all remaining trees. The impact of a harvest might be similar to the pattern of waves formed by a stone thrown into calm water. The gap caused by the thinning of a single tree will improve the growth of all adjacent trees as more resources become available. The improved growth of these trees, though, will depress the growth of their neighbors since the promoted trees can employ the additional resources for reinforcing their competition for resources with their adjacent trees, e.g. through faster or extended growth of shoots and roots. The depression of the neighbors of the neighbors to the gap will again promote their neighbors, and so forth.

Yet, the described dynamic of a thinned stand violates the assumption of a homogenous stand since the trees have accumulated different timber volumes some time after the thinning. In order to preserve homogeneity, two possible interpretations arise. First, a thinning in the homogenous stand will change the position of the remaining trees such that they will grow equally afterwards. Although this postulation surely conflicts with the observable facts, it nevertheless might serve as a heuristic approach. Alternatively, however equally conceptual, it may be assumed that the trees are growing homogenously further on despite their varying growing areas. Since the influence of a thinning on all remaining trees is equally dispersed following these interpretations, the full extent of the thinning impact is regarded as a sum while preserving homogeneity. As a consequence of this approach, the thinning intensity is an empty concept – at least for the management of a homogeneous stand and for a stand growing within a linear forest (cf. Johansson and Löfgren 1985 p. 112 ff.) – since it can never be (most) profitable to cut two trees simultaneously as all trees influence the timber volume of all other trees.

In practice, however, the effect of the harvest of a tree will not be measureable in a longer distance from the harvest gap, possibly already in the second or third ripple. Typically, the impact is assumed to be relevant on only five to seven neighboring trees. This facilitates the determination of the impact rates, which might be connected to competition indices. Hence, a second interpretation arises when the effects of harvests are supposed to operate only locally. Accordingly, condition [4-36] demands to cut as many trees such that the wavelike effects of each harvest mutually compensate resulting in a homogenous stand afterwards. This is practicable since in the homogenous stand the order of harvests is irrelevant as all trees are growing equally at equal ages before they are harvested. Consequently, different contemplated thinning intensities do not consider the same trees necessarily. Instead, alternative intensities are related to regularly distributed sampling patterns. A forest owner chooses the intensity in order to satisfy condition [4-36] while simultaneously preserving the homogeneity through balancing the impacts of all thinning gaps. As a consequence of this approach, the thinning intensity, measured as the number of trees removed, is a relevant concept as long as the stand area is large enough and/or the tree sizes are small enough to generate local effects.

Likewise, the thinning frequency, if understood as the mirror-image concept to the thinning intensity, is meaningless when all trees have different optimal harvest ages necessarily. In the case of the first interpretation of the preceding paragraphs, the thinnings would then be as frequent as possible, i.e., as numerous as the trees to be thinned. The second interpretation, on the other hand, will cause less frequent thinnings as more trees are removed at the same age.

When thinnings entail a divestment character, optimal thinning intervals are not necessarily decreasing over rising stand age

The usually applied unit areas thus ensure that thinnings capture more than one tree in almost every situation within homogeneous stands. With rising heterogeneity, by contrast, selective and frequent removals of single trees become increasingly relevant, especially in cases where fixed harvest costs are low, cf. [4-37]. Furthermore, this tendency might be growing with rising net unit revenue differentials if they do not favor the thickest trees, i.e., if they are not progressively increasing or degressively decreasing over rising diameters (cf. Section 4.4.1), as then comparatively large numbers of trees are removed in order to produce the largest diameters. Therefore, forest owners are expected to concentrate their thinnings at specific ages the more timber growth is found to be homogeneous and net unit revenues are favoring thicker trees. In the presence of commonly observed degressively rising net unit revenues, reasons for thinnings which are conducted frequently and which are distributed more or less evenly over the rotation period of even-aged stands might not often be found within an investment analysis if the fixed harvest costs are not unusually low. From this point of view, Heyer's "Golden Rule" for thinnings (1854, p. 257), i.e. early, often and moderately, might only be golden for stands which are regenerated under conditions of a comparatively high variability concerning site quality and/ or genetics, for situations of more or less price-independent timber structures and for low fixed harvest costs (presumably due to the low wage rates in these times).

In these situations, thinnings might equally be guided by the height development of the dominant trees in the stand such that thinnings are conducted in fixed intervals of dominant height growth. Here, the thinning interval necessarily increases over the rotation period as tree height growth declines (Smith et al. 1997, p. 124). This type of management, however, is only to be expected without net unit revenue differentials and more or less homogeneous growth since thinning is not only investment but also divestment (cf. Section 5.1.1). For reasons of divestment, openings in the canopy of older stands, which are not closing quickly or at all, might be acceptable as long as the removed trees were growing at unsatisfactory rates in value. The concern for openings, though, is intelligible due to the rising relative land rent, cf. [4-20] or [4-25]. If the removed trees are of comparatively low value, however, the closure of openings is not necessarily unprofitable. The same holds for progressively rising premiums for thicker trees. Then, forest owners might shorten the thinning interval, and thus thin more frequently with rising stand age.

Characterizations of the concepts of the intensity and frequency of thinnings are various. Often, the intensity is defined in terms of timber volume or basal area removed. If this is expressed as a percentage, it is equivalent to the definition in this work for homogenous stands. If it refers to an absolute amount, thinnings are then intensified for equal numbers of trees harvested at each age since the timber volume of trees increases with rising age, cf. [3-41]. The thinning frequency, on the other hand, sometimes denotes the development of the interval between two thinnings. The advantage of the approach in this work is the direct relationship between the thinning concept and the harvest ages. It thus emphasizes the relevant management implications.

5.1.3   Changes and Adaptations

Changes in the timber price and the regeneration costs lead to qualitatively equivalent changes in the Faustmann and the thinning model

The optimal thinning regime as well as its relevance is determined with the help of the investment parameters specified by the maximization approach [3-7]. As the comparative static analysis has shown, the optimal harvest ages may shift due to changes in the investment situation generated by these parameters. Within the set up Faustmann laboratory (cf. Section 3.3.1), these changes in the exogenous variables cause adaptations of the endogenous variables which might or might not be specified. The changes run from one equilibrium position to another without any proposition concerning the path they are pursuing in-between. Yet, they serve as the basis for hypotheses about the adaptions of forest owners due to changes in the incentives of timber production within the network of market exchanges.

Regarding the change in the optimal rotation age, the results of the comparative static analysis in this study (cf. Paragraph 4.5) are correlated to those of the Faustmann model (cf. Johansson and Löfgren 1985, p. 80 ff.). In either analysis, the change in the optimal rotation age due to changes in the timber prices and regeneration costs point unambiguously in the same direction. This result is based on the similarity between the *FPO* theorem [4-4] and the maximum condition in this study [4-3] as well as on the independency of the isolated optimal thinning ages on changes in these parameters. If timber prices rise, i.e., if timber becomes more valuable compared to all other goods and services provided through markets, rotation ages tend to decrease regardless of adaptations in the thinning ages. Depending on the change in the value increment at the harvest age of the remaining trees, the optimal thinning ages might equally decrease or else increase. If the timber price is separated into two timber prices, the rotation age might also increase when timber prices of thinning products rise. In general, however, if the value of timber is considered as a whole, the optimal rotation ages will decrease necessarily with rising timber prices. In some situations of rising timber prices, thinnings become less relevant if the optimal thinning ages increase while the opposite holds for decreasing optimal thinning ages. For instance, when timber prices are substituted for net unit revenues (cf. Section 4.4.2), lower variable harvest costs due to higher technical productivity, lower yield taxes on harvesting, or price supports for timber are likely to decrease the optimal rotation ages while thinnings are not necessarily conducted earlier.

The effect of rising regeneration costs is just the opposite of rising timber prices. Optimal rotation ages necessarily increase if regeneration costs rise. The impact on the optimal thinning ages, on the other hand, depends on the specifications of the impact on the increment of subsequently cut trees. Therefore, subsidies for reforestation, for instance, decrease the optimal rotation ages but do not necessarily favor earlier thinnings while missing mast years increase optimal rotation ages but might induce earlier thinnings in some situations.

The introduction of thinnings into the Faustmann model affects the unambiguity of changes in the rate of interest on the optimal rotation age

In the Faustmann model [3-2], a rise in the level of the interest rate shortens the optimal rotation age unambiguously (Amacher et al. 2009, p. 26 f.). The same does not necessarily hold for the optimal rotation age in the thinning model [3-6]. Instead, the interest rate affects both the alternative use of the tree value and the value of future harvests, however in opposite directions due to the discounting, cf. [4-13]. Propositions can only be deduced if the problem is simplified and if additional assumptions concerning the impact of the interest rate are made, cf. Table 4-1. For only two or three trees, or homogeneous tree classes respectively, the optimal rotation age might even rise with rising interest rates if the timber increments at the rotation age rise while the isolated harvest ages decrease due to a thinning. In the thinning model, this possibility cannot be excluded.

The optimal thinning ages in solitarily growing stands are independent of changes in the timber price and the regeneration costs

In general, shifts of the optimal thinning ages can only be clearly indicated when the problem is both reduced and further specified. For only two or three trees, or, equally, two or three homogenous classes of trees with equal harvest ages, the optimal thinning age adapts to the change in the same direction as the optimal rotation age if the postponement of a thinning raises the timber increment at subsequent harvests, i.e., if $q^i_{t_i t_k} > 0$, cf. [4-84] and [4-85]. In the reverse case, the change points into the opposite direction. However, the conditions under which the derivatives in [4-84] and [4-85] yield positive or negative values remain indefinite. Further model details of the timber growth theory are necessary to clarify this point.

For solitarily growing forest stands, though, the comparative derivatives can be evaluated unambiguously (cf. Table 4-1). Most interestingly, the optimal thinning ages are then independent of changes in the timber price and the regeneration costs. Since the condition for the optimal thinning age of solitarily growing trees is equivalent to the condition for a simple duration or single rotation period problem (cf. Paragraph 4.2), it is not surprising that the comparative statics yield the same results (cf. Johansson and Löfgren 1985, p. 80). Due to the simultaneous equation system, though, the impact of changes in the interest rate can only be determined with further specifications.

The comparative static analysis of Chang (1983) might be generalized to include thinnings when timber quality aspects are irrelevant

An interesting extension of the results of the comparative static analysis would be the incorporation of the initial density as an endogenous variable as in Chang (1983). However, the prospects of unambiguous propositions are low as the merging of both approaches potentiates the channels through which indirect effects of changes in the investment parameters might be transported. Naturally, if parameter changes induce higher initial densities, as for example due to rising timber prices, the optimal thinning ages will be affected. The direction of change, though, will depend on the specific magnitudes of the investment. Eventually though, thinnings might become fairly irrelevant if the initial density enters as an endogenous variable (cf. Section 4.4.3) as price differentials are exploited by modifications of the initial density (cf. Section 5.1.1). In this sense, the results of Chang (1983) might be generalized as thinnings become irrelevant.

In homogeneous stands, increases in the optimal thinning ages tend to decrease the optimal thinning intensity, and vice versa

In the model approach, the optimal thinning regime is the consequence of the optimal harvest ages. Naturally, changes in the investment parameters which are followed by changes in the optimal harvest ages lead to changes in the optimal thinning regime. The assessment of the directions of change and the conditions for their validity, however, could not be derived analytically. Therefore, only the general aspects are mentioned.

The optimal thinning intensity in the homogeneous stand is given by condition [4-36]. In principle, it is independent of the investment parameters as they can be cancelled out. Nevertheless, since $q_{t_k}^{k+1}$ is a function of the current as well as of all preceding harvest ages, it may change due to adaptions of the harvest ages. For instance, if $q_{t_k}^{k+1}$ is negative initially, and if the optimal harvest age decreases due to changes in the investment situation, $q_{t_k}^{k+1}$ might become zero when the trees share a shorter growth phase thus offering less opportunities to improve their growth by thinnings. In this way, factors leading to lower thinning ages offer incentives to thin more intense, and *vice versa*. With fixed harvest costs, this tendency is strengthened as small changes in the impacts might be absorbed (cf. Section 4.4.2.2).

In heterogeneous stands, two trees share the same optimal harvest age if condition [4-37] holds. Again, timber prices might be cancelled out if they are not separated into thinning and rotation prices. If the latter applies, rises in the thinning timber price and declines in the rotation price, respectively, directly reduce the thinning intensity as the equality in [4-37] is repealed as long as the less vigorously growing tree is cut previously. In the opposite case, the thinning will be intensified as the right hand side increases which will make it more profitable to harvest trees with differing value growth rates simultaneously. In each case, however, the interest rate remains part of [4-37]. Rising interest rates will increase the value of the differences in the future impacts of the corresponding trees. Likewise, this will tend to intensify the thinning since greater impacts allow greater differences in the value growth rates. Anyway, changes in the investment parameters give rise to changes in the harvest ages. Eventually, these indirect changes might compensate the direct changes. Analogously to the thinning intensity, the optimal thinning method is influenced by the rate of interest on the right hand side of [4-47]. With high interest rates, the right hand side is comparatively low such that thinnings from below become less relevant.

The viability of the scientific management of forest stands for profitable timber production is doubtful since, even in the simplest case, many necessary adaptions remain indefinite

Considered as a whole, the comparative statics in this analysis indicate the limits of analytical investigations of extensions of the Faustmann model. Just as in Chang (1983), even slight extensions greatly complicate the analysis as not all of the static derivatives can be determined unambiguously. On the other hand, however, there is wide conformance with the comparative static analysis of the Faustmann model. Its qualitative results thus even hold for more complex investment situations. While the impact of changes in the timber price and the regeneration cost on the optimal rotation age points into the same direction, the complex channels through which changes in the interest rate are carried do not allow clear propositions (see also Chang 1983; Li and Löfgren 2000). In this way, the assumption of the insignificance of the thinning regime on the comparative statics of the Faustmann model is supported by the presented model. The fact, however, that optimal adaptations to changing investment situations cannot be conclusively determined, even for such a simplified model as employed in this study, might illustrate the difficulty or even impossibility of scientific management. As more criteria of profitable timber production enter, such as the risk of damages, the analysis is presumably less likely to be simplified. When all derivatives become unknown, the implications become arbitrary.

5.2 Suboptimal Thinning

In contrast to the preceding Paragraph 5.1, in which thinnings where explored within the framework constructed in Section 3.3.1, the following paragraph discusses aspects of the optimal thinning regime outside of the "Faustmann laboratory" (Deegen et al. 2011, p. 363). In this broad sense, thinnings are then suboptimal as they no longer satisfy the optimality conditions derived in the analysis. Naturally, the thinning regime depends on myriads of factors. In the end, everything influences its determination. The "Faustmann laboratory", on the other hand, employs stringent analytical tools to separate a small subset of factors which satisfies the purpose to generate a solvable sub-problem while simultaneously aiming at the explanation of most of the original problem. However, the question arises in which direction the solution will change when the stringent assumptions are relaxed. In the first section, the assumption of a perfect capital market (cf. Section 3.3.1) is discussed (Section 5.2.1). The subsequent section explores the applicability of the results to a dynamic world and one of unanticipated changes (Section 5.2.2). Finally, the last section of this paragraph tries to demonstrate how heuristic approaches can be used to handle the complexity of a forest stand with respect to a management application. In this context, a heuristic approach is illustrated by means of an example (Section 5.2.3).

5.2.1   Consumption and Investment

An analytical tool employed in the Faustmann laboratory (cf. Section 3.3.1) is the conceptual separation of production from consumption. This "separation theorem" (Hirshleifer 1970, p. 63) is guaranteed by the assumption of perfect capital markets in which the rate of interest for borrowing just equals the rate for lending capital (Fisher 1930, p. 125 ff.). In this way, forest owners are able to transform any income stream from timber production to a desired one at the market rate of interest.

Naturally, perfect capital markets serve as a heuristic assumption. In observable markets, by contrast, interest rates for borrowing and lending diverge due to transaction costs (Hirshleifer et al. 2005, p. 468). Nevertheless, as long as privacy and trading rules are guaranteed (Smith 1991, p. 223), imperfect markets are efficient in the sense that the market is cleared. In this way, the market process underlying the Faustmann model ensures the adaptation of forest owners to changes in the price system (cf. Hayek 1945).

With imperfect capital markets, the investment cannot be separated from the preferences to consume since transformations of income streams via the capital market are not costless. In order to consume more today, more future income must be sacrificed than in the opposite case if the interest rate for borrowing exceeds the interest rate for lending capital. Before the implications of such borrowing constraints on the optimal thinning regime are discussed, though, the consumption of the owner of a thinned forest stand must be addressed.

The consumer choice in the Faustmann laboratory

The ultimate objective of a forest owner might be expressed by the maximization of his intertemporal utility U (cf. Section 3.3.1). If utility u is attained by consumption c, which depends in turn on calendar time t, the forest owner acts as if

$$\max_{c} U = \int_{0}^{\infty} u[c(t)] \, e^{-\delta t} dt, \qquad [5\text{-}2]$$

where δ is the subjective time preference rate of the forest owner (cf. Amacher et al. 2009, p. 41). The infinite time horizon might imply either an eternal life span of the forest owner or a forest dynasty where all heirs of the forest owner have equal preferences (cf. Salo and Tahvonen 1999, p. 107). Another important incentive, however, lies in the opportunity to sell the forested land (cf. Samuelson 1976, p. 474). In the economic equilibrium, the price obtainable for forest land equals the *LEV* which offers the greatest utility for finite life spans as well.

The opportunity of the forest owner to consume is constrained by his possibilities to earn income. Next to the income generated by timber production, the owner might receive an exogenous income $\hat{d}$ and own a wealth endowment of a_0. The forest income is given by the $LEV = LEV(t_1, \dots, t_n)$ in [3-6], i.e., forest income is generated by the regeneration, thinning and clear-cutting in periodical cycles of equal length.

Since consumption is assumed to be enabled with the help of perfect capital markets, the present value of all future consumptions is restricted by the present value of all future exchange opportunities. Hence, the intertemporal budget constraint is given by

$$\int_0^\infty c(t)\, e^{-rt} dt = \int_0^\infty \hat{d}(t)\, e^{-rt} dt + ra_0 + LEV, \qquad [5\text{-}3]$$

where r is the market interest rate. The maximization problem in [5-2] subject to [5-3] can be solved with the help of the Lagrangian function L, i.e.,

$$\max_{c, t_1, \dots, t_n} L = \int_0^\infty u[c(t)]\, e^{-\delta t} dt$$
$$+ \lambda \left[\int_0^\infty \hat{d}(t)\, e^{-rt} dt + LEV - \int_0^\infty c(t)\, e^{-rt} dt \right], \qquad [5\text{-}4]$$

where λ is a Lagrangian multiplier (cf. Amacher et al. 2009, p. 42). The optimal thinning and rotation ages are then given by the first-order condition for a maximum, which is

$$\left. \frac{\partial L}{\partial t_k} \right|_{(t_1^*, \dots, t_n^*)} = \lambda \left[pq_{t_k}^k - rpq^k + \sum_{j=k+1}^{n} pq_{t_k}^j e^{r(t_k - t_j)} \right] = 0 \qquad [5\text{-}5]$$

$$\left. \frac{\partial L}{\partial t_n} \right|_{(t_1^*, \dots, t_n^*)} = \lambda \left(pq_{t_n}^n - rpq^n - rLEV \right) = 0. \qquad [5\text{-}6]$$

Accordingly, the optimal harvest ages $t_1^*, \dots, t_n^*$ are determined irrespectively of the preferences of the forest owner since λ can be eliminated in [5-5] and [5-6] leaving an equivalent determination to [3-13] - [3-15].

In this way, forest owners can finance both consumption and investment either personally from their own endowment or by borrowing at the capital market. For perfect capital markets, lending and borrowing rates are equal such that the costs accruing from employing capital are equal. Forest owners can therefore attain their consumptive optimum, which depends on their preferences, irrespectively of their productive optimum (Hirshleifer et al. 2005, p. 468). In turn, investment actions are made independently of subjective preferences. This ensures that forest owners act as maximizing the land expectation value since the higher present values allow larger consumption sets as the ultimate objective (cf. Section 3.3.1). Naturally, if non-market utilities enter the analysis, such as *in situ*-preferences (Salo and Tahvonen 1999) or amenity services (Amacher et al. 2009, p. 45 ff.), the separation theorem does not hold since the Lagrangian multiplier, which captures the marginal utility of consumption, cannot be eliminated. In the case of linearity in the income function, the solution becomes equivalent to a multiple market use model (cf. Hartman 1976; Amacher et al. 2009, p. 72).

In this way, the presented analysis is a pure investment analysis. Forest owners invest in regeneration and (partly) in thinnings in order to provide future income. The investment is unrestricted by financial budget constraints since any amount of capital required can be borrowed from the capital market at a unique rate. At the same time, the costs for employing self-owned capital are of equal amount. Naturally, the separation theorem is an analytical tool but not an empirical observation or hypothesis. It serves to concentrate the analysis on selected issues which otherwise remain unknown or ambiguous. In the terminology of Musgrave (1981), the effective capital market serves as a heuristic assumption for the successive approximation to the problem of how people interact with respect to forests.

Optimal thinnings under borrowing constraints

Perfect capital markets are not observed anywhere. On the contrary, de-manders of current income must typically pay higher interest rates than sup-pliers due to transaction costs. Often, the borrowing rates are prohibitively high or virtually infinite when securities are missing. In this case, consump-tion of the forest owner might be financed by timber harvests at suboptimal Faustmann rotation ages rather than by borrowing capital at the market at high rates. This situation is often referred to as the "Volvo argument" (Johansson and Löfgren 1985, p. 138). For forestry, the effect of imperfect capital markets on the profitability of timber production was seminally ana-lyzed by Tahvonen et al. (2001). They demonstrate how constraints on the borrowing rate lead to significant changes in the optimal production of tim-ber and how changes in the investment parameters may lead to opposite ad-justments of the rotation age compared to the Faustmann model.

Without the intention to prove the mathematical derivations here, it is basi-cally conceivable to extend the analysis of Tahvonen et al. (2001) for differ-ent harvest ages within a rotation period such that their asset equation (3) (Tahvonen et al. 2001, p. 1600) is adjusted to lead to

$$a(t_{i,j}) = a(t_{i,j}^-) + pq^j(t_{i,1} - t_{i-1,1}, \dots, t_{i,j} - t_{i-1,j}) - C^n$$

$$i = 1,2,\dots, \quad j = 1,2,\dots,n$$

[5-7]

with

$$C^n = \begin{cases} 0 \ if \ t_{i,j} \ j = 1,2,\dots,n-1 \\ C \ if \ t_{i,j} \ j = n \end{cases},$$

[5-8]

where a is the level of financial assets at the harvest age $t_{i,j}$ of the jth tree in the ith rotation, $t_{i,j}^-$ is the left hand limit of calendar time t at $t_{i,j}$, C^n are the regeneration costs which only accrue when the last tree n is harvested, cf. [5-8]; all other notations as in Chapter 3. Thinning ages are thus separated from rotation ages by the cost incurred at the regeneration at age $t_{i,n}$.

The general optimal harvest condition (21) in Tahvonen et al. (2001, p. 1601) might then be extended to differentiate between the determination of rotation ages and thinning ages. As long as the borrowing constraint is not binding (cf. Tahvonen et al. 2001, p. 1602), the corresponding equation system might be formulated as

$$pq_{t_{i,k}}^k(s_{i,k}) - rpq^k(s_{i,k}) + \sum_{h=k+1}^{n} pq_{t_{i,k}}^h(s_{i,h})e^{r(t_{i,k}-t_{i,h})} = 0 \qquad [5\text{-}9]$$

$$\vdots \qquad\qquad \vdots \qquad\qquad \vdots$$

$$pq_{t_{i,n}}^n(s_{i,n}) - r[pq^n(s_{i,n}) - C] - r\sum_{j=1}^{n-1} pq^j(s_{i,j})e^{-rt_{i,j}}$$
$$-pq_{t_{i,n}}^n(s_{i+1,n})e^{-rt_n} = 0, \qquad [5\text{-}10]$$

where s is the stand age and $k = 1, \dots, n-1$. As Tahvonen et al. (2001, p. 1602) have proven for the Faustmann model, successive rotation ages are of equal length for perfect capital markets since the optimal rotation age is determined irrespectively of calendar time and of subjective factors regarding the forest owner such that the solution to the Faustmann model holds when the borrowing constraint is inactive. Equally, the optimal thinning regime, as explored in the analysis (cf. Chapter 4), must then hold for the equation system [5-9] to [5-10] as it is equivalent to [4-75] - [4-77].

If the subjective time preference rate δ of a forest owner exceeds the interest rate in a perfect capital market, i.e. $\delta > r$, such that the forest owner is willing to consume more today than he earns before or at the harvest ages, consumption decreases over time continuously when markets are perfect (cf. Tahvonen et al. 2001, p. 1603). Thinnings then offer opportunities to increase the LEV and thus to increase total consumption. Consumption remains continuously due to efficient market exchanges while total stand volume and financial assets are more often discontinuously interrupted when thinnings are relevant since timber value is transferred to the asset values at every harvest age. In this way, thinnings serve to expand the opportunity set of the forest owner while transformation costs are zero. The same holds for equal or lower subjective time preference rates compared to the market interest rates, however with constant or increasing consumption over time, where current income might also be lent.

With a borrowing constraint, current income might not be exchanged for future income. Forest owners with subjective time preference rates exceeding the market rate of interest are then unable to finance higher current consumption levels in exchange for future income, which is presumably generated at the harvest ages through timber production. Forest owners with lower time preference rates, on the other hand, might continue to transfer current income to future income via lending at the capital market such that the optimal thinning regime stays within the Faustmann approach. This does only apply necessarily, however, if successive rotation ages are stationary (cf. Tahvonen et al. 2001, p. 1604).

If $\delta > r$, equation system [5-9] - [5-10] does not hold for the borrowing constraint. In contrast, monetary terms are then evaluated in utility units. The optimal rotation age might change in both directions depending on the relative magnitude of the changes in the marginal revenues and costs of postponing the harvest (cf. Tahvonen et al. 2001, p. 1605). Due to the imperfect capital market, optimal consumption becomes discontinuously such that it jumps at the time of the harvest, which offers an explanation for the "Volvo argument". Depending on the asset endowment and on non-forest income, forest owners might have incentives to harvest younger trees compared to the Faustmann model where the marginal revenues of postponing the harvest might outweigh the marginal costs of holding the land and timber value at the market rate of interest whereas the postponement does not account for the comparatively high costs due to the subjective time preference rate.

In this case, thinnings offer opportunities to finance consumption while simultaneously allowing for further revenues from holding the forest stand. As long as the cost-covering diameter is exceeded, consumption can be financed by thinnings. The alternative of employing the tree value for consumption, i.e. the first term on right hand side of [4-13], is then more valuable than the transformation to the capital market. This tends to reduce the optimal thinning age, or it renders thinnings profitable at all. Even without impacts on the remaining trees, thinnings might be conducted to obtain consumption opportunities. On the one hand, this will reduce the consumption possibilities at the rotation age as the stand value is reduced. On the other hand, however, harvest prior to the rotation period might finance consumption on a level above the non-forest income. Together, both effects tend to level out the optimal consumption path over time. Naturally, this balancing effect is constrained by the number of trees, the land rent and fixed harvest costs. In contrast to uneven-aged management, thinnings are thus unable to smooth out consumption entirely, but they surely allow for consumption above the non-forest income before the rotation age.

With non-stationary rotation ages (cf. Tahvonen et al. 2001, p. 1612), the initial assets of the forest owner help determine the convergence process towards the stationary rotations of the Faustmann model, or the constrained borrowing model respectively, if mining of the forest stand within one or a few rotation periods as well as regeneration delays are excluded. Only a sufficiently high initial level of assets of forest owners with lower subjective time preference rates than the market interest rate ($\delta < r$) guarantees the optimality of the Faustmann solution and thus of the thinning model presented in this study. If the initial assets are lower than the net timber value at the rotation age (i.e., the timber value at the rotation age minus the regeneration costs), on the other hand, the Faustmann and thinning solution are only stationary states in a finite long run. The current rotation age is then shorter since the convergence is from below (cf. Tahvonen et al. 2001, p. 1616). In the same way, thinnings within the current rotation might be antedated in order to finance consumption at an earlier time even for forest owners with incentives to save. Conversely, forest owners with high initial assets and high time preference rates have incentives to harvest at the beginning according to the Faustmann rule and the thinning rule, respectively. Not until the initially high assets are more or less depleted, rotation and thinning ages are reduced to finance consumption.

For stationary rotation ages, Tahvonen et al. (2001, p. 1606 ff.) were able to comparatively statically analyze the effect of a borrowing constraint. The effect of rising regeneration costs and timber prices are analogous to those within the Faustmann model (cf. Amacher et al. 2009, p. 26 ff.) although exceptions exists for the timber price effect. The impact of the interest rate on the optimal rotation age, on the other hand, follows a non-monotonic course depending on the subjective time preference rate. Although not comparable directly, these results are analogous to the comparative static analysis in this study where the effect of the timber price and the regeneration costs on the optimal rotation age is equal to those within the Faustmann model framework.

In summary, thinnings might offer an important source for financing consumption when capital markets are not perfect. As long as the cost-covering diameter is exceeded, harvests prior to the rotation age might finance consumption of forest owners with comparatively high subjective time preference rates and/ or low initial endowments. This tends to increase the relevance of thinnings and its share on the total amount of timber harvested. The marginal revenues of satisfying consumptive demands might thus increase the marginal costs of holding the tree. Precommercial thinnings, on the other hand, are ruled out if the borrowing constraint is active, i.e. for comparatively high time preference rates or low initial endowments, as the forest owner is unable to invest on credit.

5.2.2   Dynamics, Risk and Uncertainty

Viewed from the perspective of the cyclic harvest and regeneration of trees, the approach in the analysis might be termed static as all future states are assumed to be constant (cf. Section 3.3.1). Within each cycle, though, the development of the timber volumes of the trees follows a dynamic course as the timber volume is different at different ages. Moreover, the movement between static equilibria due to exogenous changes generates dynamic hypotheses since the disequilibrating cause must occur within a time period (cf. Section 3.3.1). In this sense, dynamics is any internal movement within the model. With reservations, then, the thinning model [3-6] can be generalized to be applied to a dynamic world, although the conceptuality of statics and dynamics is varying among different authors (cf. Machlup 1959).

Furthermore, the approach in this study is deterministic as random events are assumed not to occur. In a stochastic setting, by contrast, specific events may occur with a given probability. If these events are anticipated, rational forest owners respond by adjustments depending on the expected probability of occurrence.

Finally, the analysis in this study is based on certainty as all future stand states are assumed to be known by the forest owner. Again, this assumption serves as a heuristic device since the future is always uncertain in the sense that neither the probabilities of occurrence nor the possible future stand states are known (cf. Knight 1921, p. 19 ff.). Instead, unanticipated changes can be observed to occur constantly. In this way, the preceding analysis concentrates on optimal ages as opposed to optimal points in (calendar) time.

Changes over the rotation ages

The parameters and variables in the thinning model [3-6] might be subject to change over the rotation periods. In general, if the net value V^n at the end of the rotation period t_n, i.e. $V^n = \sum_{i=1}^{n} pq^i e^{r(t_n - t_i)} - Ce^{rt_n}$, changes at a rate of g from period to period, the LEV^g with changes over the rotation ages follows the sequence

$$LEV^g = \frac{V^n e^{g0}}{e^{rt_n}} + \frac{V^n e^{gt_n}}{e^{r2t_n}} + \frac{V^n e^{g2t_n}}{e^{r3t_n}} + \cdots = \sum_{i=1}^{\infty} \frac{V^n e^{g(i-1)t_n}}{e^{rit_n}}$$

$$= \frac{V^n}{e^{rt_n} - e^{gt_n}}.$$

[5-11]

Accordingly, changes over the rotation ages are basically induced via the capitalization sequence in the denominator of the LEV. If the rate of change applies to the first rotation already, LEV^g has to be scaled by e^{gt_n}, which equals $V^n/(e^{(r-g)t_n} - 1)$ (cf. Johansson and Löfgren 1985, p. 108).

In the Faustmann model [3-2] as well as in the thinning model [3-6], the "pesky little '-1'" (Gaffney 2008, p. 124) indicates the static character over the rotation ages as all future timber values and regeneration costs are assumed to be of equal magnitude, i.e. $g = 0$. If future net values are expected to increase, i.e. $g > 0$, the *LEV* increases since $e^{gt_n} > 1$. However, if the rate of net value change equals or exceeds the rate of interest, i.e., if $g \geq r$, the final equality in [5-11] does not hold since the sequence does not converge, or it is not defined respectively. In this case, the *LEV* would be infinite. Since infinite wealth is not observed, the problem must be approached differently. One solution is suggested by Johansson and Löfgren (1985, p. 108 ff.).

If future net values are expected to decrease, such that $g < 0$, the *LEV* decreases since $e^{gt_n} < 1$. As $g \to -\infty$, the *LEV* approaches V^n/e^{rt_n} which gives the one rotation age or duration solution (cf. Hirshleifer 1970, p. 82ff). The latter could be interpreted as the "worst" case scenario (Lu and Chang 1996, p. 284) of value decline.

The necessary conditions for a maximum of the LEV^g are in general

$$\frac{\partial LEV^g}{\partial t_k}\bigg|_{(t_i^*,\ldots,t_n^*)} = pq_{t_k}^k e^{-r(t_n-t_k)} - rpq^k e^{-r(t_n-t_k)}$$

$$+ \sum_{j=k+1}^{n} pq_{t_k}^j e^{-r(t_n-t_j)} = 0 \qquad [5\text{-}12]$$

$$\frac{\partial LEV^g}{\partial t_n}\bigg|_{(t_i^*,\ldots,t_n^*)} = pq_{t_n}^n - (r+g)pq^n - (re^{gT} - ge^{rT})LEV^g = 0. \quad [5\text{-}13]$$

Accordingly, the optimal harvest ages remain stationary as marginal costs and revenues remain equivalent in different rotation periods due to the constant rate of change. The isolated determination of the optimal thinning ages remains unchanged as [5-12] is qualitatively equivalent to [3-14]. Consequently, the basic determination of the optimal thinning regime (cf. Paragraph 4.3) applies in principle. The optimal rotation age, though, decreases for $r > g > 0$ (and increases for $g < 0$) as, compared to [3-15], the cost side

increases unilaterally due to higher costs for postponing higher future revenues. Changes in the rotation age affect the optimal thinning ages by virtue
of the simultaneous equation system. Hence, the relevance of thinnings is affected (cf. Paragraph 4.2) such potentially fewer trees are thinned when the
rotation age decreases, and *vice versa*.

With a constant rate of value change over the rotation ages, the general analysis in this study remains unchanged. Although the optimal thinning regime
differs with value changes over the rotation period, the optimal harvest ages
remain stationary such that the maximum conditions only have to be adjusted for potential rates of change. Naturally, different components of the
net value at the rotation age could be subject to different rates of change.
Despite the complication of the determination of the optimal harvest ages,
the basic analysis, however, remains valid.

The sources of value changes are diverse. An important source is inflation,
i.e., a rise in the general price level of all goods and services, which tends to
decrease the rotation ages and thus the relevance of thinnings. In this sense,
g can be interpreted as the "Teuerungszuwachs", i.e. the rate of price increment (Chang and Deegen 2011, p. 259), which Pressler (1860) identified as
the third source of value increment (cf. Section 4.4.1). As another source, timber growth might improve (e.g. Löfgren 1985) or deteriorate (e.g. Halbritter
and Deegen 2011) over the rotation periods. When future timber values decrease, for instance due to exploitations of the site productivity, the optimal
rotation period increases and thinnings become potentially more relevant as
they offer the opportunity to utilize the current rotation period more intensive.

Stochastic thinnings

When forest owners anticipate different future stand states and expect them
to occur with different probabilities, the optimal cutting regime of the forest
stand is influenced. Typically, forest owners must face various aspects of
risk. These might be timber price (Brazee and Bulte 2000), interest rate or
timber growth risks (cf. Amacher et al. 2009, p. 241). One important kind of

risks is the occurrence of catastrophic events, such as fire, pests, wind or snow damage or drought (cf. Gardiner and Quine 2000). While some of these might be influenced by the management, others occur randomly. Here, only basic aspects of randomly occurring catastrophic events are considered in order to indicate the general impact on the optimal thinning regime. Hence, the probability of occurrence is assumed to follow a homogeneous Poisson process (cf. Amacher et al. 2009, p. 269), where the arrival of the event is constant over time.

As Reed (1984) has shown, the Faustmann model might be transformed to a stochastic version for randomly and totally destructive catastrophes, which destroy the total timber value of the current stand. In this case, a forest stand is afflicted at the average rate of catastrophic occurrence per time unit γ as the Poisson parameter. If it is assumed that the timber value is worthless after the catastrophe and the stand has to be regenerated for regeneration costs in the amount of C_2, the LEV with risk of random catastrophic events and thinnings might be written as (cf. Reed 1984, p. 184)

$$
LEV^r = r\left(1 - e^{-(r+\gamma)t_n}\right)^{-1}\left[(r+\gamma)\left(\sum_{i=1}^{n} pq^i e^{-(r+\gamma)t_i}\right.\right.
$$

$$
\left.\left. - Ce^{-(r+\gamma)t_n}\right)\right] - \frac{\gamma}{r}C_2. \qquad [5\text{-}14]
$$

Maximizing LEV^r with respect to the harvest ages yields

$$
\left.\frac{\partial LEV^r}{\partial t_k}\right|_{(t_i^*,\dots,t_n^*)} = pq_{t_k}^k e^{-(r+\gamma)t_k} - rpq^k e^{-(r+\gamma)t_k}
$$

$$
+ \sum_{j=k+1}^{n} pq_{t_k}^j e^{-(r+\gamma)t_j} = 0 \qquad [5\text{-}15]
$$

$$
\left.\frac{\partial LEV^r}{\partial t_n}\right|_{(t_i^*,\dots,t_n^*)} = pq_{t_n}^n
$$

$$
-\left(1 - e^{-(r+\gamma)t_n}\right)^{-1}\left[(r+\gamma)\left(\sum_{i=1}^{n} pq^i e^{-(r+\gamma)(t_n-t_i)} - C\right)\right] = 0. \qquad [5\text{-}16]
$$

In the face of catastrophic events, the optimal rotation age decreases since the average rate of occurrence enters the maximum condition via the discount factor thus increasing the costs of holding the stand unilaterally (Reed 1984, p. 184). In the same way, the optimal thinning ages are expected to decrease since the present value of the impact on the remaining trees, i.e. the sum on the very right of [5-15], is reduced. With the decreasing present value of the competition impact, the relevance of thinnings (cf. Paragraph 4.2) decreases while the thinning intensity (cf. Section 4.3.1) increases and the thinnings tend to be conducted from below (cf. Section 4.3.1). Similar results might be obtained with the expected utility approach for risk-averse forest owners (cf. Johansson and Löfgren 1985, p. 271).

The approach might be extended for age-dependent arrival rates, where the Poisson parameter changes over the age. Naturally, with a monotonically increasing average rate of occurrence over the stand age (e.g. in the face of storm threats), the optimal harvest ages further decrease (cf. Reed 1984, p. 190) since the present value of the remaining trees is reduced. In cases, in which catastrophic events do not lead to the total destruction of the standing timber value, the LEV^r in [5-14] might be extended by the present value of the mean randomly salvageable timber value (cf. Reed 1984, p. 188). Since the latter is increasing over the harvest ages if timber production is profitable, the marginal costs of holding the trees rises thus increasing the harvest ages compared to the totally destructive event.

More important for thinnings is the possibility to influence the average arrival rate of catastrophic events by management efforts (cf. Reed 1987; Reed and Apaloo 1991). In this case, the thinning model [3-6] offers an interesting opportunity to analyze the impact of thinnings on the remaining trees. One possible way might be to assume that the Poisson parameter of a tree i is dependent on its age as well as on already conducted harvests prior to the rotation age, i.e., $\gamma^i = \gamma^i(t_1, ..., t_i)$. In the face of storm risk, for instance, $\partial \gamma^i(t_1, ..., t_i)/\partial t_j < 0$ with $j < i$ if the stability of individual trees rises due to increasing stem diameters. In this case, more trees tend to be thinned (and from above, cf. Section 4.3.4) as the present value of the remaining trees increases due decreasing threats of storm damage. On the other hand, if

$\partial \gamma^i(t_1, \ldots, t_i)/\partial t_j > 0$ with $j < i$ due to a decrease in the collective stability of stand by virtue of gaps, the opposite applies. In this way, the change in the Poisson parameter might be adjusted for the specific risk in case.

Many other kinds of risks might enter the thinning model. For instance, the initial density might be adjusted when some trees are expected to be dying off randomly during the stand development due to microsite-specific causes such as browsing or fungal or insect infestation. On the one hand, this might induce forest owners to regenerate more trees if the expected loss from fewer trees justifies the additional regeneration costs (cf. Section 4.4.3). If, then, fewer trees drop out during the stand development than expected, thinnings become more relevant since the competitive pressure is higher than expected. Without the opportunity to thin and thus to adjust the investment if more trees are present than expected, forest owners might not invest in more trees at the beginning of the rotation period since the losses due to possibly thinner trees at the rotation age must be considered. However, thinnings might equally be unprofitable such fewer trees are regenerated. The complexity of the maximization problem increases progressively by virtue of the simultaneous equation system. As a consequence, seemingly reasonable responses of forest owners might become ambiguous, such as the direction of the adjustment of the initial density in the face of mortality risks.

Occurrence of unanticipated changes

The LEV in the thinning model [3-6] refers to the value of bare land. Forest stands, however, might also comprise timber volume already. The forest value FV of such a forest then differs from the LEV. If the standing timber is assumed to be clear-cut in some time followed by instantaneous regeneration and subsequent harvests in infinite cycles, future rotation ages can be determined independently of standing timber in the same way as in [3-7] due to Bellman's (1957) principle of optimality, according to which every part of the optimal path must be optimal (cf. Johansson and Löfgren 1985, p. 86). With standing timber, the maximization problem might be given by

$$\max_{t_i, \dots, t_n} FV = \sum_{i \leq n}^{n} pq^i(t_1, \dots, t_i)e^{-r(t_i - t_a)} \tag{5-17}$$
$$+ LEV(t_1^*, \dots, t_n^*)e^{-r(t_n - t_a)},$$

where $t_1^*, \dots, t_n^*$ are the optimal harvest ages while t_a is the current age of the stand. The maximum conditions for the optimal harvest ages are

$$\frac{\partial FV}{\partial t_k}\bigg|_{(t_i^*, \dots, t_n^*)} = pq_{t_k}^k e^{-r(t_k - t_a)} - rpq^k e^{-r(t_k - t_a)}$$
$$+ \sum_{j=k+1}^{n} pq_{t_k}^j e^{-r(t_j - t_a)} = 0 \tag{5-18}$$

$$\frac{\partial FV}{\partial t_n}\bigg|_{(t_i^*, \dots, t_n^*)} = pq_{t_n}^n e^{-r(t_n - t_a)} - rpq^n e^{-r(t_n - t_a)}$$
$$- rLEV e^{-r(t_n - t_a)} = 0, \tag{5-19}$$

for $k \geq i$. Since the stand age t_a can be eliminated in both [5-18] and [5-19], the optimal harvest ages are independent of the standing timber volume as [5-18] and [5-19] are equivalent to [3-13] - [3-15]. The land value LV with standing timber is then simply $LEVe^{rt_a}$ (cf. Johansson and Löfgren 1985, p. 86).

The independence of the optimal harvest ages in [5-18] and [5-19], however, does only remain valid if the standing timber is produced along the optimal path determined by the current investment situation. In an uncertain world, by contrast, the standing timber value might be given by

$$\sum_{f=1}^{l} pq^f(t_1, \dots, t_f)e^{-rt_f}, \tag{5-20}$$

where $q^f(t_1, \dots, t_f) \neq q^i(t_1^*, \dots, t_i^*)$ since future developments could not be assessed correctly. The present value PV of the land enclosing this timber volume might be termed "holding value" (Klemperer 1996, p. 222). The maximization problem and the corresponding first-order conditions are

$$\max_{t_1,\dots,t_l} PV = \sum_{f=1}^{l} pq^f(t_1,\dots,t_f)e^{-r(t_f-t_a)}$$
$$+ LEV(t_1{}^*,\dots,t_i{}^*)e^{-r(t_l-t_a)} \tag{5-21}$$

$$\left.\frac{\partial PV}{\partial t_u}\right|_{(t_1^+,\dots,t_s^+)} = pq_{t_u}^u e^{-r(t_u-t_a)} - rpq^u e^{-r(t_u-t_a)}$$
$$+ \sum_{j=u+1}^{l} pq_{t_u}^j e^{-r(t_j-t_a)} = 0 \tag{5-22}$$

$$\left.\frac{\partial PV}{\partial t_l}\right|_{(t_1^+,\dots,t_s^+)} = pq_{t_l}^l e^{-r(t_l-t_a)} - rpq^l e^{-r(t_l-t_a)}$$
$$- rLEVe^{-r(t_l-t_a)} = 0, \tag{5-23}$$

where $t_u \geq t_a$. Although the discount factors can be eliminated, the optimal harvest ages of the current rotation $t_1^+, \dots, t_l^+$ do not necessarily coincide with the optimal harvest ages within the Faustmann approach $t_1{}^*, \dots, t_i{}^*$. If the timber volumes differ, i.e. if $q^f(t_1, \dots, t_f) \neq q^i(t_1{}^*, \dots, t_i{}^*)$, the optimal harvest ages differ and also the comparative static analysis (cf. Chang 1998, p. 656).

Naturally, if the optimal management at a given point in time and for a given forest stand according to [3-7] does not comprise thinnings, they might be conducted if the timber volume of the current forest stand diverges from the optimal timber volume. If, for instance, the stand has evolved out of a high initial density due to the expectation of high prices for thin trees, thinnings might be conducted which have not been anticipated in order to produce thicker trees in shorter time if the timber price for those trees has risen since then. Or *vice versa*, initially intended thinnings of thin trees might be postponed when the price for timber of high quality is comparatively high.

From this point of view, thinnings offer the opportunity to adjust the investment in timber production to unanticipated changes. For instance, if expectations in high prices for thin trees have been disappointed, thinnings offer the opportunity to adjust the timber investment into the direction of thicker trees. In a stationary economy with constant or constantly changing price levels, thinnings might often be irrelevant if exceptionally high premiums for timber quality are excluded (cf. Section 4.4.3). The incentives to regenerate trees for thinnings are then very low since the price structure is already regarded by the initial density and additional timber volumes cannot be produced. It can thus be expected that thinnings are the less likely to be conducted the more constant the expectations of the emergence of prices progresses. Such a stationary order, however, can only evolve with dynamic adaptations on the part of the individuals involved since some changes in the conditions surrounding human actions occur which demand adjustments if conditions remain stable. It is just these adjustments to unanticipated changes which render economic actions necessary (Hayek 2002).

Unanticipated changes in the conditions individuals face necessarily result in the disappointment of the expectations of some individuals (Hayek 1973). If thinnings, then, are anticipated to be profitable, unanticipated changes might render them unprofitable, and *vice versa*. The relevance of thinnings (cf. Paragraph 4.2) is directly influenced by the new circumstances surrounding the forest owner since the new optimal harvesting regime enters the current regime via the *LEV* in the equation system [5-22] - [5-23]. New unanticipated changes in the future will again change the optimal management and thus the current thinning regime. If unanticipated changes are then observed to be occurring constantly, the thinning regime of forest owners is a "permanent process" (Deegen 2001, p. 35) of adjustments on the basis of the presented model. Therefore, observed forest stands in market economies are expected to be the result of all adjustments to past changes in the price levels and not the single maximization solution on the basis of past or current price levels. How often and unanticipated investment situations change has been illustrated for forestry by Raup (1966). Whenever the direction of change deviates from the direction expected from the comparative static analysis (cf. Paragraph 4.5), those action conditions might be traced which caused the deviation.

The adjustment character of thinnings in an uncertain world becomes even more evident if thinnings are understood as the reverse operation to increasing the initial density. In this way, they allow to adjust the initial investment in the initial density as thinnings are *ex post* modifications of the initial density (cf. Section 5.1.1). Therefore, thinnings qualitatively influence the timber volume and its structure in the same way as reductions of the initial density while postponements of thinnings have the same impact as an increase in the initial density (cf. Chapter 2). Only the intensity of the impact of thinnings is less than the impact of changes in the initial density since the period in which the trees are able to respond to the changed situation is shorter and older trees might in general be less responsive. If the expectations of future price levels at the age of the regeneration have been disappointed, forest owners have the opportunity to adjust the initial density via thinnings. At large, if the investment conditions between the regeneration and the current age have changed, forest owners are expected to adjust to the new situation by thinnings or by postponing intended thinnings.

With uncertainty however, arbitrariness enters the analysis. If only future prices are insinuated to be high or low enough, any management regime can in principle be justified. For instance, if future price premiums for high quality timber are taken to be exceptionally high, initially very dense stands and frequent thinnings will eventually become profitable despite current price levels. In the same way, all efforts to minimize biotic or abiotic damages are justified if the future loss in volume or quality is only evaluated high enough regardless of the probability of the occurrence of the damage. Arbitrariness, though, leads to metaphysical propositions as the set of falsifiability becomes empty (Popper 2002b, p. 68 ff.). In this case, the scientific argument goes the other way around: forest owners observed to regenerate their forest stands comparatively densely are expected to have high expectations concerning quality price premiums, or *vice versa*.

As a consequence, discussions on the optimal solution to the cutting regime in forest stands might be structured by the separation of diverging opinions concerning the known and the unknown effects on future states. On the one hand, ambiguities arise due to the complexity of the problem (cf. also

Puettmann et al. 2009) for a certain future with known changes (cf. Table 4-1). On the other hand, different opinions result from the complex or simple, but unknown future. Disputes over current thinning regimes can thus be viewed as disputes over causal relationships and future market conditions. Albeit these assumptions are not always stated, they are nevertheless implied. In this way, debates can be analyzed in order to "expose them [faiths] to the light of inquiry" (Duerr and Duerr 1975, p. 41).

5.2.3   Heuristics and Adaptive Management

The system of first order conditions [3-13] - [3-15] required to be satisfied simultaneously consist of n equations, i.e., one condition for each tree. Applied to actual forest stands, the system might comprise 500 to 10,000 or more equations. In view of this variety, any complete solution is virtually impracticable. It is not just that all the required information is not obtainable, but also that the data is only static in the model laboratory. As for general economics, simultaneous equation systems can only serve as a tool for the main problem, but not seriously as the solution itself (Hayek 1945). The equation system thus serves as an analytical tool for generating hypotheses in a precisely defined environment.

For the case of an application with empirical data, heuristics have to be employed which reduce the analytical set to a practicable format. For this, any heterogeneous stand might be assumed to be comprised of two or more classes of trees which compete for resources, i.e.

$$\sum_{i=1}^{n} pq^i(t_1, \dots, t_i) \equiv \sum_{j=1}^{k} pQ^j(t_1, \dots, t_j) \quad with \; k < n, \qquad [5\text{-}24]$$

such that n trees are reduced to k classes with Q as the timber volume of a class. These classes are assumed to be growing homogeneously as well as independently enough to form a non-competitive collective with a uniform harvest age (cf. Section 4.2.2). Naturally, this approach bears resemblance to the optimal thinning intensity (cf. Section 4.3.1) as the trees are assumed to have the same optimal harvest age. The assumption of more or less independent growth within each tree class is practicable since the classes are intensively intermingled such that trees of the same class are more or less spatially separated from each other. The number of equations is then reduced to the number of classes selected, whereby the comparative static analysis has already been made controllable (cf. Paragraph 4.5). The number of classes to be selected depends on the situation to be examined. In some situations, for instance for low *LEV*s or low levels of competitive pressure, it might suffice to separate only two classes.

The separation of forest stands into tree classes is related to the concept of cohorts in general resource economics (cf. Beverton and Holt 1957, p. 135 ff.; Getz and Haight 1989, p. 136 ff.; Wacker and Blank 1998, p. 88 ff.; Clark 2005, p. 217 ff.). In contrast to the latter, though, the separation of forest stands into tree classes might be viewed as an effort to integrate density-dependent structures into the age-based Faustmann model while cohorts are typically used to integrate age-dependent structures into density-dependent fishery models, as Deegen (2002) pointed out. However, the classification of the trees of a stand serves as a heuristic approach to handle the complex varieties of forest stands just as cohorts should for fishery.

The relevant criterion for the separation of tree classes within this analysis is the value growth rate (cf. Section 4.3.4). Only if trees growing independently of each other share equal value growth rates, they share the same optimal harvest age (cf. Section 4.2.2). This might be extended to account for fixed harvest cost (cf. Section 4.4.2.2). With a unique timber price for all timber volumes, the criterion reduces to the timber volume growth rate as the timber price is cancelled out. If thinning concepts contemplate thinnings within a tree class, i.e., if only parts of a tree class are to be removed, the assumption of independent and/ or homogeneous growth within a class does not hold. In this case, the classification does not serve as a heuristic for the determination of the trees to be cut since thinnings within each class have to be assessed as well.

In forestry, tree classification systems for organizing trees into groups with similar features are commonly applied, at least since Kraft (1884). This classification system is also used to define silvicultural thinning concepts (cf. Smith et al. 1997, p. 99 ff.). As introduced in Section 4.3.4, the Kraft (1884) classes might be interpreted to serve the objective to classify trees according to their timber volume growth rates. Interpreted in this sense, the Kraft (1884) classification becomes both relevant at all as well as heuristically convenient for the management of forest stands. On the one hand, if the Kraft (1884) classes order trees according to different criteria (e.g. timber volume increments), it would be useless for the assessment of profitable timber production as qualitatively equal investment situations would remove trees of different classes. Hence, the trees to be removed could not be named unambiguously. On the other hand, the Kraft (1884) classes serve as an efficacious tool for heuristic approaches of the stand management. Despite (or because of) their simplicity, they are easily applicable and allow to focus on the relevant aspects of timber production: the crown indicates the timber increment (due to the pipe model theory, cf. Section 2.1.2), the stem indicates the timber volume; set in relation, the volume growth rate is revealed.

Often, the Kraft (1884) classification system has been criticized for its lack of quality criteria. In this course, other classification schemes, such as the IU-FRO or Assmann (1970) system, have been developed to account for this deficiency. In a way, these efforts might be interpreted as an approximation towards value growth rates as the relevant determinants of the optimal harvest ages since both classifications are only equivalent if the net unit revenue for timber is unique. However, the decisive advantage of the classification according to Kraft (1884) lies in its conceptual separation between the less and more predictable speculation variables. While tree growth in the near future might often be comparatively well assessed, investment situations might totally change just the other day. The speculative momentum does thus lie in the simulation of future market conditions. Even in the face of climate change, market conditions as well as the preferences of forest owners often change more rapidly and on a smaller scale than the characteristics of forest sites. Quality criteria, which have been relevant today, might be less important or irrelevant tomorrow. In this situation, the classification system today is inappropriate for tomorrow. The volume growth rates, on the other hand, are relevant for timber production in general independent of time and location. Consequently, the Kraft (1884) classes allow speculating on the more volatile variables of profitable timber production. On its basis, additional and more individual criteria might then be introduced, as Pressler (1860) has shown with the quality and the price increment rates, which might be extended by impact rates.

Naturally, the evolution of different tree classes within the stand development is equivalent to the evolution of trees within a heterogeneous stand (cf. Paragraph 2.2). Depending on the genetics, the site characteristics and the initial state of the trees, the highest classes of even-aged stands dominates at the beginning of the stand development as most trees are growing solitarily (Oliver and Larson 1996, p. 148 ff.). In the subsequent competition phase, more vigorously growing trees are able to reinforce their competitive advantage such that inferior trees fall behind more and more, i.e., reduce their growth rates. This leads to a transition of trees to lower classes.

While the origins of different growth rates are rooted in genetic and site dissimilarities (Assmann 1970, p. 83), the relevant discrepancies between growth rates are the result of mutual interdependencies since the concept of the forest stand implies more or less similar growth conditions. In this way, thinnings become relevant through mutual interdependencies between trees in the heterogeneous stand as well since differing value growth rates (or rather their large differences) are the result of competition. The additional incentive in heterogeneous stands (cf. Section 4.2.2) is hence given by bygone interdependencies which led to the difference in the growth rates. Moreover, differing growth rates indicate density-dependent mortality. Thinnings are then conducted in order to anticipate mortality. The evolution of the tree classes is influenced by thinnings. When a thinning removes a tree class, this class will, in most cases, only be eliminated at the age of the thinning. Since growth and competition will continue after the thinning, the eliminated class will be build up again. Merely when the remaining stand is fairly sparse or it is harvested altogether soon after the thinning, the eliminated class will not be accumulated once more.

This rebuilding of the tree classes, though, complicates the heuristic approach. If the class is built up again after a thinning, the growth function of the class becomes discontinuous. One solution to this problem might be piecewise differentiation which, however, amounts merely to the separation of two classes or more which are temporally separated. In this way, equal Kraft (1884) classes might be cut at different ages which denote different classes in the model approach of this work. The concept of tree classes becomes thus equivalent to the thinning intensity since the latter combines all trees of equal growth rates which are growing independently of each other. In either way, the two-stage process of the determination of the optimal cutting regime remains valid (cf. Section 5.1.1).

In order to illustrate the working of the thinning model numerically, the described heuristics have to be employed. For simplicity, a forest stand is analyzed where a thinning reduces the stem number by 10% of the initial density by definition. For a heterogeneous stand, this could be equivalent to the removal of one or more tree classes; for a homogeneous stand, this might simply refer to the thinning intensity. These intensities or classes are exogenously given to the example, possibly by prior optimization in the face of operational restrictions (cf. Sections 4.3.1 and 4.4.2.2).

The optimal harvest ages are determined by the marginal approach which compares marginal revenues and costs on the basis of the maximum conditions [3-13] - [3-15]. This solution technique is chosen in order to demonstrate an adaptive management approach that allows the stepwise adjustment of the growth conditions to the subjective situation of a forest owner. As Chang and Deegen (2011) have shown, this approach is feasible for the determination of the rotation age in a world of unanticipated changes. Accordingly, the relevant changes in the investment as specified by the first order conditions are condensed in terms of an indicator rate which allows speculating about the attainment of the optimal harvest age in a simple way. With reference to the optimal thinning ages, this approach has the advantage to estimate the impact of a thinning on the value of the remaining trees as a simple percentage change since the required data are seldom available.

Table 5-1 displays the illustration example. It has been constructed based on a frequently employed stand timber volume function Q for Loblolly pine (*Pinus taeda* L.) from Chang (1984):

$$Q(T) = exp\left(9.75 - \frac{3,418.11}{m * T} - \frac{740.82}{T * si} - \frac{34.01}{T^2} - \frac{1,527.67}{si^2}\right), \quad [5\text{-}25]$$

where the site index si has been set to 60 and the initial density m to 607.04, which corresponds to 1,500 pines ha^{-1}. All values in Table 5-1 have been converted to m^3 and ha. In connection with exogenously given constant prices and costs, columns 2 to 4 illustrate the determination of the LEV according to the Faustmann model [3-1] and [4-2]. The optimal Faustmann rotation age t_a is then approximately 18 years since the LEV in column 4 is there at a maximum of about $431 ha^{-1}.

Columns 5 to 8 demonstrate the relevant criteria for the optimal thinning regime. Since the stand is assumed to grow homogenously, all living trees share equal growth rates, which are displayed in column 5. These growth rates differ from the rates given by the growth model [5-25] when thinnings modify the timber volume and its increment. As the net unit revenues are constant in this example, value and volume growth rates are of equal magnitudes. The potential to influence the timber volume of the remaining trees is included in column 6, which is termed the thinning indicator rate (TIR). The latter is defined as

$$TIR := \sum_{j=k}^{n} \frac{pq_{t_k}^{j}}{pq^k} e^{r(t_k - t_j)}, \quad [5\text{-}26]$$

which is obtained by adding up all impact rates on one side of [4-14]. The relative impact of a thinning on the remaining trees is thus given by subtracting the growth rate from the TIR.

Table 5-1 The optimal Faustmann rotation age and the optimal harvest ages of the thinning model

1	2	3	4	5	6	7	8
age [years]	Q [$m^3 ha^{-1}$]	Q_t [$m^3 ha^{-1}a^{-1}$]	LEV^s [$\$ ha^{-1}$]	$pq^i_{t_i}/pq^i$ [a^{-1}]	TIR^A [a^{-1}]	PIR^B [a^{-1}]	LEV [$\$ ha^{-1}$]
5	5.53	6.98	-2,156.94	1.263	0.802	-0.046	-2,156.94
6	15.26	12.42	-1,632.57	0.814	0.457	-0.100	-1,632.57
7	30.07	17.00	-1,173.35	0.565	0.274	-0.214	-1,117.35
8	48.78	20.18	-782.10	0.414	0.168	-0.721	-782.10
9	70.01	22.07	-460.26	0.315	0.103	0.942	-460.26
10	92.58	22.94	-203.68	0.248	*0.060	0.346	-203.68
11	115.65	23.09	-4.69	0.213	0.128	0.214	-4.59
12	138.61	22.76	145.50	0.173	0.095	0.156	147.13
13	161.07	22.12	255.45	0.144	0.072	0.124	259.99
14	182.79	21.30	332.79	0.129	*0.062	0.109	341.43
15	203.63	20.37	384.04	0.111	0.080	0.093	398.54
16	223.52	19.41	414.61	0.097	0.067	0.081	436.84
17	242.45	18.44	428.92	0.089	*0.060	0.075	460.26
18	260.41	17.49	*430.54	0.080	0.066	0.067	472.49
19	277.43	16.57	422.35	0.074	*0.060	*0.062	*476.31
20	293.56	15.69	406.65	0.067	0.061	0.057	473.87
21	308.83	14.86	385.28	0.062	0.056	0.054	453.99
22	323.29	14.07	359.70	0.058	0.051	0.051	444.72

$r = 0.06\ p.\,a., p = \$10\ m^{-3}, C = \$600\ ha^{-1}$, A Thinning indicator rate, B Pressler's indicator rate

Due to missing data and for the sake of simplicity, the thinning impact on the remaining trees has been freely estimated as well as reduced to the impact rate on the trees harvested at the rotation age by the function

$$\frac{q^n_{t_k}}{q^k} = -17.153t^{-1.725}0.5^x, \tag{5-27}$$

where t is the stand age and x is the number of thinnings already conducted. Accordingly, it is assumed that the impact rate of a thinning follows a decreasing course at a decreasing rate over the stand age. This assessment is based on the circumstance that the potential impact on the timber volume of the remaining trees decreases over the age as the phase of competition is prolonged. The decline is more or less rapid as the response to thinnings of older trees is typically low. At the same time, the tree value of the thinned trees increases as it accumulates more timber volume. Combined as in [5-27], the impact rate thus decreases at a decreasing rate similar to the course of the growth rate. The diminution of the potential to influence the remaining trees after thinnings have already been conducted is ensured by a reduction factor raised to the power of the number of thinnings conducted.

The *TIR* is qualitatively analogous to Pressler's indicator rate (*PIR*). The latter is defined as (cf. Pressler 1860)

$$PIR := \left(\frac{pq_{t_n}^n}{pq^n}\right) \frac{(n-k)pq^n/LEV}{(n-k)pq^n/LEV + 1}. \qquad \text{[5-28]}$$

[5-28] is attained by isolating the interest rate in [4-19]. For the example in Table 5-1, the course of *PIR* is shown in column 7. Here, the quality and price development is held constant. According to the *PIR* criterion (cf. Gong and Löfgren 2010), the optimal rotation age is reached when *PIR* is equal to the interest rate provided the intersection is from above. Equally, an optimal thinning age is reached when the *TIR* is equal to the interest rate. Since the potential to influence the remaining trees after a thinning has been conducted decreases, the *TIR* criterion can be met multiply.

In the example above, the *TIR* criterion is satisfied at the ages of 10, 14, 17 and 19 years. These are the optimal thinning ages indicated by the stars in the column. At these ages, 10% of the initial trees are removed, i.e., an assumed homogenous class of 150 ha^{-1} trees at every thinning. The last optimal thinning age, though, coincides with the optimal rotation age as *PIR* in column 7 becomes equal to the interest rate. Here, the *LEV* is at a maximum of approximately \$476 ha^{-1}, which is \$46 ha^{-1} or 10.6% more than the maximum of the Faustmann model in column 4. As a result, the optimal thinning regime comprises three thinnings with a total of 450 ha^{-1} trees removed prior to the rotation age.

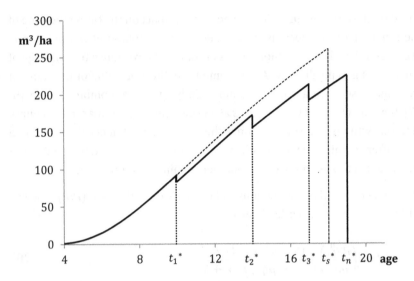

Figure 5.2 The timber volume of the untreated and thinned stand in the example

In this example, many simplifications have been made, in particular regarding the impact rate of a thinning. However, the basic approach can be illustrated. At any age, the forest owner assesses with the help of *PIR* whether the optimal rotation age has been reached by speculations about the development of value growth rates and land rents. If the final harvest is postponed, the forest owner assesses the satisfaction of the *TIR* by estimating future value growth and impact rates for different thinning intensities. If the *TIR* is equal or less than the interest rate, the stand is thinned; otherwise any harvests are postponed to the next period, in which the same procedure is repeated until *PIR* is finally satisfied, the stand is clear cut and reforested. Naturally, the approach in the example can be specified to account for the different sources of value growth for both *PIR* and the *TIR* (cf. Chang and Deegen 2011).

Figure 5.2 shows the development of the timber volume of the stand in the example. While the dashed curve represents the timber volume without any treatments, the solid curve reveals the timber volume development of the thinned stand. The three thinnings cause the characteristic notches of thinned forest stands, where 150 pine trees ha^{-1} are removed. Since older pines comprise more timber volume, the thinning intensity, when measured as the timber volume removed, increases with the stand age. The intervals between the thinnings, on the other hand, decrease. As explored in Section 4.3.3, both the development of the impact rates and the growth rates determine the optimal thinning interval. In this case, the interval decreases since the mining character of the thinning, i.e. the right hand side of [4-45], prevails. Nevertheless, due to the exogenously determined intensity, later thinnings could be more intense for an unconstraint maximum which might in turn lengthen later thinning intervals.

Eventually, Figure 5.2 illustrates the advantage of thinnings in a homogeneous stand clearly. Each thinning reduces the timber volume of the stand until the next harvest while simultaneously maintaining or slightly reducing the timber increment (cf. Section 4.4.4). Therefore, the growth rates increase as well as the *LEV* since equal or slightly less output is produced with less input. However, the timber increment is not increased since in this case it pays to thin more intensively in order to further reduce the timber volume such that even higher growth rates are obtained. As a result of rising growth rates, the optimal rotation age might likewise increase. This proposition, though, cannot be generalized as the land rent increases simultaneously which tends to shorten the rotation period (cf. Section 4.1.1.1).

5.3 Thinnings and Interactions

The present study is a contribution to the action theory of forestry (cf. Chapter 1). However, the basic problem of forestry economics is the cooperation of individuals for mutual advantages. For this object of investigation, action theory serves as the microfoundation for the individual action before cooperation (or conflict respectively) takes place; i.e., the actions of other individuals are taken as exogenously given restriction. The actual decisive interaction theory, on the other hand, explains the coordination of different individual actions. At this level, the forest owner appears as a bargaining supplier offering timber and capital for exchange. Naturally, thinning is an important strategic tool in this context as it allows for flexibility. Here, only a few implications are addressed, i.e., the influence of thinnings on the timber supply (Section 5.3.1), on the distribution and allocation of income and land with respect to forestry (Section 5.3.2), and on the formation of forests by market conditions (Section 5.3.3).

5.3.1   Individual Timber Supply

In the presented model, the optimal harvest ages determine the individual timber supply, which is defined as the timber volume per time unit the forest owner is willing to sell in a perfect market at a specific timber price. In this way, the optimal timber volume of tree q^{i^*} is determined by the optimal harvest ages such that

$$q^{i^*} = q^{i^*}(t_1^*, \dots, t_i^*). \qquad [5\text{-}29]$$

Since the optimal harvest ages are determined with the help of the investment parameters, it equally holds that

$$q^{i^*} = q^{i^*}[t_1^*(z), \dots, t_i^*(z)], \qquad [5\text{-}30]$$

where the second-order condition [3-16] is assumed to hold and z is a vector comprising the relevant investment parameters, i.e. in this case, $z = p, C, r$. For the entire forest stand of n trees, the optimal timber volume is then

$$\sum_{i=1}^{n} q^i[t_1^*(z), \dots, t_i^*(z)]. \qquad [5\text{-}31]$$

[5-31] defines the optimal total growth performance of the stand within a rotation period. In each rotation period, q^{i^*} is produced at age t_i^*.

Though [5-31] refers to different harvest ages, each tree is cut at equal rotation intervals, i.e., every t_n^* years, due to the stationary rotation ages (cf. Section 5.2.2). In order to compare the timber output in different investment situations, the total timber growth performance in [5-31] is annualized over the rotation period, i.e.,

$$s^T \equiv \frac{\sum_{i=1}^{n} q^i[t_1^*(z), \dots, t_i^*(z)]}{t_n^*(z)}, \qquad\qquad [5\text{-}32]$$

which typically defines the short-run timber supply s^T (cf. Amacher et al. 2009, p. 28; Conrad 2010, p. 145).

This concept of the individual, annual timber supply, however, is bound to the concept of a linear (Johansson and Löfgren 1985, p. 112) and normal or fully regulated forest (Binkley 1993) where all age classes up to the optimal rotation age are present with the same share of the total forest area A, i.e. A/t_n^*. Only in this case, the forest owner is willing to supply the share of the optimal total growth performance annually. Moreover, it must be assumed that changes in the rotation age are accompanied by instantaneous changes in the age class distribution to a normal forest without any conversion costs. In this way, the individual timber supply in the Faustmann model does only serve as a preliminary to an actual microeconomic timber supply. Important parameters for the determination of the microeconomic timber supply are not comprised in the Faustmann model as a basically stand level approach (cf. Section 5.4.3 for more discussion).

From the comparative static analysis (cf. Paragraph 4.5), the short-run timber supply in [5-32] can be derived directly. Accordingly, the timber supply decreases with rising timber prices and increases with rising regeneration costs due to decreasing and increasing optimal rotation ages (cf. Table 4-1) and monotonically rising timber volumes, cf. [3-41]. The impact of the interest rate, on the other hand, remains ambiguous. In order to evaluate the long-run timber supply, [5-32] must be differentiated with respect to the investment variable for assessing the impact of variable parameters. Therefore,

$$\frac{ds^T}{dz} = \frac{t_n^* \sum_{i=1}^n \sum_{j=1}^i q_{t_j}^i (t_j^*)_z - (t_n^*)_z \sum_{i=1}^n q^i}{(t_n^*)^2}$$

$$= \frac{t_n^* \sum_{i=1}^n \sum_{j=1}^{i<n} q_{t_j}^i (t_j^*)_z + t_n^* q_{t_n}^n (t_n^*)_z - (t_n^*)_z \sum_{i=1}^n q^i}{(t_n^*)^2} \qquad [5\text{-}33]$$

$$= \frac{(t_n^*)_z}{t_n^*} \left[\sum_{i=1}^n \sum_{j=1}^{i<n} q_{t_j}^i \frac{(t_j^*)_z}{(t_n^*)_z} - \frac{\sum_{i=1}^{n-1} q^i}{t_n^*} + q_{t_n}^n - \frac{q^n}{t_n^*} \right],$$

where $(t_i^*)_z = dt_i^*/dz$. Without the first two terms inside the bracket, [5-33] is qualitatively equivalent the timber supply of the Faustmann model (Amacher et al. 2009, p. 29). The third and the forth term within the bracket cause the supply function to be backward-bending (Hyde 1980, p. 67) since the difference is positive for rotation ages shorter than the age of the maximum sustained yield and negative in the opposite case (Binkley 1987).

With thinnings, the first two terms inside the bracket are nonzero. The first term sums all impacts on the remaining trees as well as the timber increments of the thinned trees. Together with the second term, the thinning influence on the timber supply can be written as

$$\sum_{i=1}^n \sum_{j=1}^{i<n} q_{t_j}^i \frac{(t_j^*)_z}{(t_n^*)_z} - \frac{\sum_{i=1}^{n-1} q^i}{t_n^*}$$

$$= \sum_{i=2}^n \sum_{j=1}^{i-1<n} q_{t_j}^i \frac{(t_i^*)_z}{(t_n^*)_z} + \sum_{j=1}^{n-1} q_{t_j}^j \frac{(t_j^*)_z}{(t_n^*)_z} - \frac{\sum_{i=1}^{n-1} q^i}{t_n^*}. \qquad [5\text{-}34]$$

The first term on the right hand side denotes all additional impacts of trees on remaining trees which are adjusted for the ratio between the optimal thinning and rotation age. In a solely competitive forest stand, this term is negative since all impacts are negative or zero, cf. [3-47]. The second and third term on the right hand side are necessarily positive as all trees have positive timber volumes and increments, cf. [3-41]. While the timber volumes are annualized over the rotation age, the increments are adjusted for the ratio between the optimal thinning and rotation age. The sign of the adjustment factor cannot be specified in general. If the cross partial derivative of the timber volume is positive, the adjustment factor is positive for two or three trees or tree classes necessarily (and presumably also for more trees or classes) since the change in the harvest ages with respect to the timber price and the regeneration costs points in the same direction (cf. Table 4-1). In the opposite case, the adjustment factor is negative. The sign of changes in the interest rate remains indefinite.

Combined as in [5-34], the net impact on the individual timber supply is thus ambiguous. For positive adjustment factors, if the sum of the adjusted increments outweighs the prorated timber volumes and the thinning impacts, the timber supply will increase over a longer range as in the Faustmann model. In the opposite case, as well as for negative adjustment factors, the timber supply will be backward-bending already for lower timber prices.

The backward-bending of the long-run timber supply derived of the Faustmann model is contradicting the monotonically increasing supply functions in microeconomic models. Often, this feature is ascribed to the limited availability of land and its limited productivity. Certainly, the Faustmann supply disregards the opportunity to increase the long-run timber supply by allocating more land to forestry as well as the opportunity to intensify timber production on the existing areas. The latter effect might partly be explained by changes in the initial density (Hyde 1980; Chang 1983) and partly by changes in the thinning regime. If the timber price rises, it might pay to plant more trees thus increasing the timber supply. With regards to thinnings, however, more or less timber might be supplied in either direction of the change in the investment situation as the change in the optimal thinning ages remains indefinite.

5.3.2   Allocation and Distribution

The allocation of land to forests has two different dimensions (Deegen et al. 2011). The temporal allocation of forest lands is the balancing of total land available between forestry and non-forest uses. The intertemporal forest land allocation, on the other hand, is the division of the total forest area for current and future consumption, i.e., for current and future timber harvests. Naturally, thinnings influence both dimensions of land allocation as the optimal thinning ages are important determinants of the forest land rent.

The first implications are straightforward. If thinnings are relevant, they increase the *LEV* thus lifting the competitiveness of forestry compared with other land uses as the intersection point of competing land rents is shifted (Strand 1969). In this way, more land of the total spectrum is allocated to forestry other things being equal (Amacher et al. 2008). Without competitive land uses, the extensive margin of forest land use (Zhang and Pearse 2011, p. 170 f.) is extended as some previously idle land generates positive land rents (Hyde 1980, p. 73 ff.). Typically, forestry is a capital-intensive and labor-extensive land use compared to other forms, such as agriculture. Thinnings influence this capital-/labor-relation in a particular manner. Since forestry is capital-intensive due to the long periods of timber production, relevant thinnings reduce the accruing capital costs as the capital stock is lowered in relation to the value increments (cf. Section 4.4.4). On the other hand, as thinnings require the employment of labor, the labor intensity of the timber management is increased since thinner trees cause higher harvest costs (cf. Section 4.4.2.1). Therefore, shifts in the labor/ capital cost relation between forestry and non-forest land uses will be less pronounced when thinnings are relevant. When thinnings are precommercial, on the other hand, additional capital and labor have to be supplied.

When an increase in the quantity of labor employed is connected to the operation of thinnings, the harvest of trees prior to the rotation period is a shift of the intensive margin of land use (Zhang and Pearse 2011, p. 169) towards intensified management on each land unit. Together, the extensive and the intensive margin of forest land use determine the total production of timber.

The intertemporal allocation of forest land, in turn, is in principle determined by the division of the total forest area by the rotation age (cf. Deegen et al. 2011, p. 355) which defines the consumption of timber today and in the future. Since the impact of thinnings on the optimal Faustmann rotation period is ambiguous (cf. Section 4.1.1.1), the change in the area consumed today and hereafter depends on the magnitudes of the changes in the marginal revenues and costs. Since the comparative static derivatives of the Faustmann model are qualitatively equivalent to those of the thinning model with the exception of the interest rate (cf. Paragraph 4.5), the changes in the intertemporal land allocation due to changes in the investment conditions point in the same direction, whereas the impact of the interest rate, however, remains indefinite. Additionally though, effects of changes in the age class distribution have to be considered (Salo and Tahvonen 2004).

Finally, the distribution of the income generated with timber production among the owners of the land, the capital and the labor employed (Deegen et al. 2011, p. 356 f.) is influenced by thinnings. With (commercial) thinnings, capital income on the labor and land invested is partially reduced as the harvest age of some trees decreases as long as the LEV and the rotation age remain unchanged. In this case, the land income is unaffected. The income of the laborer increases since thinner trees cause higher harvest costs (cf. Section 4.4.2.1). If thinnings increase the LEV, on the other hand, the total income of the land owner increases since forestry is more profitable. Nevertheless, the impact of thinnings on the optimal Faustmann rotation age is ambiguous (cf. Section 4.1.1.1). Therefore, with thinnings, more capital income and less labor income might be generated if the rotation age increases. This will affect the relative proportion of the incomes on the total income generated by timber production. Lastly, with profitable precommercial thinnings, more labor, capital and land income is generated. Since the increase in the total income generated is then divided between all owners of the factors of production, land owners have incentives to avoid precommercial thinnings in order to gain a larger share of the income.

5.3.3   Silviculture and Markets

As has been discussed in Section 5.1.1, silvicultural treatments can be interpreted as responses to the investment situation generated by the market conditions. Although the market conditions are the consequence of the interactions of individuals, they are exogenously given for the forest owners if their own actions in a market society are not significant due to their negligibly small fraction on the total exchanges within the market. Thus, forest owners simply react to changes in the restrictions generated by all individuals. The presented model might help to explain those changes in the silvicultural treatments which are caused by market exchanges.

For instance, as revealed by the changes in experimental plots in Section 2.3, planting densities have severely declined in the course of the last decades for many tree species in Central Europe. In view of the thinning model [3-6] and the timber growth theory (Figure 2.1), high planting densities can only be explained by low cost-covering diameters (cf. Section 4.4.3). Only for tree species with quality criteria favored by intensive competition, higher initial planting densities are to be expected. As in the calculation examples of von Thünen (1875, 2009), low wage rates at these times might have led to very low cost-covering diameters which allowed to exploit the idle land between the small trees at the beginning of the rotation period. Without the opportunity to thin, though, the initial densities would be considerable lower in order to guarantee the desired stem diameters of sawlogs in a shorter time. Without the expectation of profitable sawlog production, on the other hand, the stands would have been clear-cut in young ages. In these cases, and in combination with the attempt to reduce the regeneration costs and to increase the initial density, tree stumps were left to regenerate new and many shoots. In these coppice forests, thinnings have often been unprofitable since even thinner stems are not worth the high harvest costs. When the productivity of the stumps declined, however, new trees have been planted (Medema and Lyon 1985; Tait 1986). If promising to be suitable, then, some of these trees have been retained for sawlog production. For analyses of these two-aged stands, the thinning model would have to be adjusted (cf. Section 5.4.1 and 5.4.2).

In the 20th century, the cost-covering diameter has risen, possibly due to decreasing demand for small trees as firewood. If quality criteria are then not predominating, the optimal initial density decreases as thinning in young ages would only generate additional costs (cf. Section 4.4.3). However, at the end of the 20th century, the prevalence of fully mechanized logging and machining technologies held down the cost-covering diameter again. For Norway spruce (*Picea abies* L.) in Central Europe, the differences in the cost-covering diameter have been sometimes as large as 40 years for equal initial densities; on the other hand, they are not as low as at the times of von Thünen.

Furthermore, the changes in the machining technology resulted in the vanishing of price premiums for already thick trees. This severely reduced the incentives to invest in already thick trees as only reductions in the variable harvest costs remain attractive. In this case, higher initial densities and increased thinnings become relevant as the growing area per tree decreases in value. Moreover, thinnings tend to be conducted more from below as the net unit revenues tend to be constant over some diameter ranges (cf. Section 4.4.1.2).

In many intensively managed forest plantation in the world, thinnings are observed to be less important as most of these forests are never thinned. On the one hand, this might be explained with reference to the greater degree of homogeneity in forest plantations. In the presence of fixed harvest costs, then, thinnings become unprofitable (cf. Section 4.4.2.2). Furthermore, rotation periods become increasingly shorter such that changes in the market conditions might solely be met with adjustments of the initial density. Only in regions with low wage rates and low fixed harvest costs, thinnings might prove to be profitable as an intensified management.

In temperate and boreal latitudes, on the other hand, rotation periods are often long enough such that changes in the market conditions shape forests in various ways. In connection with the different opportunities of forest owners to invest their capital, the dynamics of market exchanges give rise to quite heterogeneous forest, even if composed of the same tree species. This development leads to the simultaneous coexistence of dense and sparse and old and young stands with thick and thin trees. In this context, many opportunities for increasing the profitability with silvicultural treatments arise.

5.4   Stands and Forests

The preceding analysis and discussion has focused on the stand as the basic management unit of forests. These stands have been separated into homogeneous and heterogeneous stands, implicitly composed of one tree species. The thinning model [3-6], however, is basically of a greater generality. The following sections discuss its employment as a model of mixed and multiple-use stands (Section 5.4.1), of uneven-aged stands (Section 5.4.2) as well as a plain model for the analysis of entire forests (Section 5.4.3).

5.4.1   The Mixed and the Multiple-Use Stand

The presented thinning model [3-6] is suitable for the analysis of both homogeneous and heterogeneous stands (cf. Paragraph 4.2). The latter are distinguished by potentially deviating value growth rates of the trees forming the stand. No reference, however, is necessary for the origin of the dissimilar value growth. In this sense, the deviations might be caused by genetic or site inequalities (cf. Section 5.2.3). For the economic analysis, though, it is irrelevant which kind of genetic differences are observed in a forest stand such that the differences might be intra- or interspecific. As a consequence, heterogeneity concerning the value growth rates might equally refer to different tree species mixed in the same stand. In this way, the results of the analysis in Chapter 4 do also apply to even-aged mixed stands.

A simple example might be constructed with the hypothetical mixture of two tree species with differing optimal Faustmann rotation ages, such as Norway spruce (*Picea abies* Karst.) and European beech (*Fagus sylvatica* L.) or grand fir (*Abies grandis* Lindl.) and Douglas fir (*Pseudotsuga menziesii* Franco). The optimal harvest ages of the combined production within an even-aged stand (i.e., regeneration is postponed until the last tree is cut) are determined by the optimal rotation age [4-3] for the tree species with the longer Faustmann rotation age and by the optimal thinning age [4-14] for the tree species with the shorter Faustmann rotation age. If inter- and intraspecific competition is disregarded, the two tree species form two independently growing tree classes according to [5-24] with deviating growth patterns. In this case, the optimal harvest age of the tree species with the shorter Faustmann rotation age will rise compared to the pure stand since the relative land rent does not have to be borne, cf. [4-29]. The harvest age of the other tree species will decrease in turn as the land rent of the earlier harvested tree species has to be compensated since [4-22] holds for solitarily growing stands. Hence, the optimal harvest ages converge and might eventually coincide in the mixed stand.

With mutual interdependencies between the trees and the species, the convergence process is modified. Considering interspecific competition, the optimal harvest ages will diverge again as the impact rate in [4-14] becomes relevant. The change in the optimal rotation age of the mixed stand, though, is ambiguous (cf. Section 4.1.1.1). Eventually, with intraspecific competition and heterogeneous growth among the trees of the same species, the optimal harvest ages might mix such that poorly growing trees of the tree species with the longer isolated rotation period are cut before well growing trees of the other species.

However, in mixed forest stands, even positive interdependencies between the volume growths of different tree species might be observed. While trees in pure stands necessarily compete for the same ecological niche, different tree species in mixed stands might occupy different ecological niches such that mutual reinforcing timber growth might possibly occur. For instance, these situations might be relevant where the one species reduces abiotic or biotic threats of the other species. With mutually reinforcing timber growth, the optimal harvest ages of the tree species in mixed stands tend to converge (cf. also Section 5.1.2).

With either inter- or intraspecific competition, another comparison might be drawn between the relative land rent and the impact rate (cf. Section 5.1.1).

The relative land rent in the *FPO* theorem [4-4] and the rotation age condition [4-3] are typically associated with the costs of holding the land value (cf. Johansson and Löfgren 1985, p. 80) since in a partial equilibrium the land value is defined as the infinite income stream generated by periodic regeneration and harvests of forest stands. In the same way as the impact rate of the optimal thinning age condition [4-14] might be interpreted as the costs of regeneration (cf. Section 5.1.1), it might equally be understood as the costs of holding the land which is associated with the thinned tree, i.e., its growing area. Since every standing tree occupies land that could be employed alternatively, it incurs costs in the amount of the next best alternative use. The next best alternative of the growing area of a tree ready for thinning is the additional timber value which might be produced by its neighboring trees when the thinning is conducted. The land that is freed from the thinned trees is thus not without use after the thinning is conducted, but, instead, it enables an additional value increment of the remaining trees, which can only be attained by keeping the adjacent land, i.e., the potential growing area, tree-free. The utilization of the land of a thinning gap can be illustrated by the poor or even non-existing growth of a small tree planted into the gap.

This kind of interpretation is closely related to combined land use concepts where different parts of the area are used for different products. Again, these multiple-use concepts simply represent heterogeneous stands where different goods or services are produced simultaneously. In either way, the intensification of the management of one crop will influence the yield of the other crop. Whether these different crops are different tree species or trees and crops or trees and livestock or trees and some services is unimportant for the general economic consequences, i.e., each intensified utilization bears costs to the owner. In the presented model, these costs emerge as the second term on the right hand side of [4-13]. Without this distinction, both the optimal harvest ages and the land use values are overrated since the marginal costs on the right hand sides of [4-3] and [4-13] are reduced. In this instance, the conclusion applies only to even-aged multiple-use stands. However, as discussed in subsequent Section 5.4.2, the results might be generalized.

Due to its employment as a multiple-use model, the thinning model [3-6] is closely related to other multiple-use models. The most prominent model might be the Hartman model (cf. Hartman 1976; Strang 1983; Amacher et al. 2008, p. 43 ff.). If thinnings are interpreted as a continuous flow of timber dependent on the rotation age and if the mutual independencies are removed, both models are equivalent. Hence, in contrast to the Hartman (1976) model, the thinning model as a multiple-use model regards the mutual interdependencies between the different products which might then lead to different optimal termination ages of the different products. In this way, the focus is more on the economically relevant mutually exclusive aspects of simultaneous production. Nevertheless, some extensions of the Hartman (1976) model incorporated some of these aspects (e.g. Swallow et al. 1990; Swallow and Wear 1993; Koskela and Ollikainen 2001; Li and Löfgren 2000).

In either case, it should be emphasized that these multiple-use models are only meaningful for goods and services exchanged in markets. Since the underlying Faustmann model [3-2] is a market model employing the Fisherian separation theorem (cf. Section 5.2.1), non-market goods and services cannot be integrated without the loss of economic significance. Without this "network of relationships that emerges or evolves out of this trading process" (Buchanan 1964, p. 220), subjective preferences might not be separated from the investment decision. One solution to this problem is proposed by Salo and Tahvonen (1999).

5.4.2   The Uneven-Aged Stand

When the removal of a tree prior to the rotation age is followed by regeneration, the resulting stand is uneven-aged. In this way, thinnings are a concept of the even-aged stand if they are understood as harvests without subsequent regeneration. Since the Faustmann model [3-2] is defined for an even-aged stand as all living trees are of the same age necessarily, thinnings are a concept of the Faustmann model. Viewed from a different angle, though, the Faustmann model combines both uneven-aged characteristics by allowing to regenerate every patch of land after a tree has been harvested as well as it comprises thinning aspects through the uniform regeneration after the final harvest. This synthesis is guaranteed by the equal harvest ages of each tree.

As indicated in Section 3.2.2, the thinning model [3-6] might technically be transformed to a more general many-aged model [3-38], where the even-aged stand represents a special case. In this way, the restriction on the timber volumes, which excludes any regeneration of the stand before the harvest of the last tree, is eliminated. The harvest of a tree might then be followed by instantaneous regeneration of the patch of bare land a tree leaves behind. Since, in this case, any trees might be influenced by the harvest of any other tree due to mixed ages, [3-39] applies. In this way, each tree is taken as constituting a single stand.

For a correct analysis, though, it must be assumed that the stand in question is already on the optimal path of the timber development such that the current stand structure already represents the result of the optimal harvest ages of each tree in the current investment situation, as in Chang (1981) or Hall (1983). This ensures that the incentives in each cutting cycle of each tree are equivalent such that cutting cycles are stationary (cf. Tahvonen et al. 2001, p. 1602). In this case, the optimal harvest age of a tree is given by setting [3-40] to zero, i.e., after rearranging

$$pq_{t_k}^k = rpq^k + rLEV^k - \sum_{i \in (n|i \neq k)} pq_{t_k}^i \frac{e^{rt_k} - 1}{e^{rt_i} - 1}.$$ [5-35]

Hence, each tree is optimally cut according to an extended form of the *FPO* theorem [4-2]. Naturally, the value increment and the capital cost on the tree value have to be considered as well as the cost of the land per tree characterized by its *LEV*, i.e. $LEV^k = (1 - e^{-rt_k})^{-1}(pq^k e^{-rt_k} - C_k)$. The latter represents the costs of regenerating the tree, i.e., to repeat its timber growth infinitely. The selection of the unit area is irrelevant as long as it applies to all trees considered. Only if LEV^k is equivalent for each cutting cycle such that $LEV^k = LEV^k(t_1^*, \dots, t_n^*)$, the harvest cycles are stationary and the uneven-aged model [3-38] is meaningful.

Next to the determinants in the *FPO* theorem [4-2], the impacts on the value of all other trees influenced by the harvest are relevant for the determination of the harvest ages in the uneven-aged stand, cf. [5-35]. They are adjusted for potentially varying cutting cycles of the trees via a capitalization sequence adjustment factor $(e^{rt_k} - 1)/(e^{rt_i} - 1)$. If the impacts are negative due to competition, they appear as costs such that the optimal harvest ages of trees in an uneven-aged stand are shorter compared to their solitary management. Time lags between the harvest and the regeneration are here implicitly regarded in the growth functions, or in the regeneration costs respectively. The chosen unity of the LEV^u in [3-38] and the LEV^k in [5-35] is open to the observer of the forest as long as it is equal for all *LEV*s. In this way, problems arising from varying growing areas over the age of a tree, which might evolve in tree-based approaches, are excluded by the selection of a unit area for LEV^u in [3-38] and the summing of the LEV^k of the same unit area of all trees which might potentially grow within the cutting cycles.

In the Faustmann model [3-2], and the thinning model [3-6] respectively, the position on the optimal path of the timber value is conventionally guaranteed by the unified regeneration age. In either case, the *LEV* is defined for bare forest land. Necessarily, the bare land represents points on the optimal path of the development of the timber volume or value as they are independent of all possible stand states since the optimal rotation age and the optimal regeneration age coincide. Thus, if a stand state is optimal at the age $t_i > 0$, its optimal rotation age coincides with the optimal rotation age of the Faustmann model (Johansson and Löfgren 1985, p. 86) while they diverge in the opposite case. On this account, the Faustmann solution emerges as a steady-state solution in an optimal control setting (Anderson 1976). In the uneven-aged stand, by contrast, there is no unified point like the bare land which is equal for all possible stand states.

As a consequence, the solution to the combined thinning and rotation problem is qualitatively analogous to the solution to the optimal harvest ages in the uneven-aged stand. Any tree is cut according to its value growth rate adjusted for the influence on all successively cut trees. Whether the influence is exerted on equally old and equally growing trees or on smaller and younger trees or even on not already existing trees is irrelevant for the general economic consequences. In this way, all analytical results equally apply to uneven-aged stands provided they are growing on the optimal path. Naturally, the latter assumption disregards any transition processes towards the optimal stand states (cf. Tahvonen 2009; Tahvonen et al. 2010). Nevertheless, and analogously to Section 5.2.2, these dynamic aspects do not render the validity of the static results. In an uncertain world of unanticipated changes, dynamic transition processes are as constantly changing as the maxima in the static approach (cf. Section 5.2.2), which, in either way, demands for heuristic and adaptive applications (cf. Section 5.2.3). Hence, dynamic problem formulations of the uneven-aged stand (cf. Haight 1985) are just as suitable as static approaches for uncertain investment situations. They differ in the path to the solution, but not in their applicability to the observable world.

In the view of this study, uneven-aged management is simply the breaking down of the stand to its smallest unit, the tree, while accounting for the interdependencies between these basic stand units. The same approach might be applied to the even-aged stand of the Faustmann model [3-2] and the thinning model [3-6]. Without interdependencies and with homogeneous growth, the optimal harvest ages in the Faustmann model are equal such that the LEVs for trees are simply the proportional share of the LEV of the stand, i.e. $LEV^k = LEV/n$. It is convenient to combine several trees with similar harvest ages (due to heterogeneous growth) to stands with equal harvest ages (i.e. the Faustmann model) as potential management costs, such as fixed harvest costs (cf. Section 4.4.2.2), administrative costs or supply-induced price impacts, might be reduced more sharply than the loss from the deviations of the single optimal harvest ages of each tree.

With interdependencies, the combination of trees might likewise be convenient (thinning model) as long as the costs of a unification of the regeneration age measured in the postponed regeneration of some bare patches of land are less than the management costs of smaller stand units. With positive interdependencies, on the other hand, the system moves towards even-aged management as the optimal harvest ages converge (cf. Paragraph 4.2).

Therefore, without any management costs, forests are managed on the basis of the smallest units, i.e. one tree. In the face of site and genetic differences as the origin of heterogeneous growth and/ or of competition between the trees, uneven-aged management will then inevitably arise as harvest and regeneration ages of different trees will differ necessarily (cf. Section 4.2.2).

5.4.3   Forest Stands and Forests

The Faustmann model [3-2] might be interpreted as to be constructed for analyses on the forest stand level as the basic management unit with sufficiently uniform conditions. The combination of different stands might then be referred to as a forest. In this sense, the application of the Faustmann model to forests is invalid when these different levels of observation deviate in their characteristics. As Johansson and Löfgren (1985, p. 112 ff.) have shown, the Faustmann model can be applied to forests constrained by a "linear technology" (Johansson and Löfgren 1985, p. 114). This linear forest might be exemplified as a normal or fully-regulated forest (Amacher et al. 2009, p. 213 ff.) where the normal age class distribution (cf. Section 5.3.1) is ensured regardless of changes in the optimal rotation ages (Salo and Tahvonen 2002).

Likewise, though, the linear forest might be defined as the sum of totally independently growing stands. By analogy, then, the thinning model [3-6] might be interpreted as constituting a nonlinear forest with interdependencies between the stands. In this way, trees are substituted for stands in which the trees are growing homogenously and/ or independently enough to share equal harvest ages (cf. Paragraph 4.2). Since each stand might be regenerated instantaneously after it is clear-cut, the more general many-aged model [3-38] applies such that the forest value FV, which is understood as the LEV of an entire forest comprising several stands, is simply the sum of the interdependent land expectation values of each stand, i.e.

$$FV = \sum_{i=1}^{n} LEV^i(t_1, \ldots, t_n)$$

$$= \sum_{i=1}^{n} (1 - e^{-rt_i})^{-1}[pQ^i(t_1, \ldots, t_n)e^{-rt_i} - C_i], \qquad [5\text{-}36]$$

with Q^i as the timber volume of the ith stand, which could be further specified as the sum of the trees, and t_i as its corresponding rotation age. As in the Faustmann model [3-2] and the thinning model [3-6], the LEV as the net present values of the bare land of the stand ensures that the initial state is on the optimal path (cf. Section 5.4.2).

The general first order maximum condition for the kth stand is then

$$\left.\frac{\partial FV}{\partial t_k}\right|_{(t_1^*,\ldots,t_n^*)} = \frac{\partial LEV^k}{\partial t_k} + \sum_{j=1 \neq k}^{n} \frac{\partial LEV^j}{\partial t_k} = 0. \qquad [5\text{-}37]$$

The first term of the derivative gives the FPO theorem [4-2] of the kth stand. However, the stand within the forest is only cut at the optimal Faustmann rotation age if the second term is zero. The latter denotes the present value of all influences on the value of the other stands of the forest. Without interdependencies, the impact is zero, and each stand within the forest is cut at its optimal Faustmann rotation age. With nonzero impacts though, however, the optimal interdependent rotation age is obtained at a different age. With predominantly negative impacts on the values of its neighboring stands, the optimal rotation age decreases; with predominantly positive impacts, the optimal rotation age will rise.

Hence, in the presence of interdependencies, the problem of the cutting cycle within a forest is qualitatively equivalent to the cutting cycle of both uneven-aged and even-aged stands (cf. Section 5.4.2 and Paragraph 4.3). It follows that the analytical results of this study might be applied to each case. Thus, as the optimal thinning regime (cf. Paragraph 4.3) is the consequence of the optimal harvest age of each tree within the stand, the optimal age structure of a forest results from the optimal rotation age of each stand. The age class distribution is thus not the primary optimization objective but an indirect consequence of harvest decisions.

The interdependencies between the stands in a forest might be manifold. As in the uneven- or even-aged stand (cf. Section 5.4.2 and Paragraph 4.3), the stands in a forest might be interlinked by biotic and abiotic factors. For instance, the harvest of a stand might increase the risk of storm damages or sunburns or reduce the threat of fires or pests of neighboring stands. Considering forests, though, the distinction between timber volume and value (cf. Paragraph 4.4) becomes even more important. If several stands in a forest have reached the optimal Faustmann rotation age, the short-run increase in the timber supply might cause the timber price to decline in the corresponding region when the capacities on the timber demand side are exhausted. In this way, the harvest of a stand reduces the unit net revenue of the next stand to be cut thus postponing its harvest. If, on the other hand, higher logging volumes promise price premiums, more stands are cut simultaneously due to positive interdependencies between the stand values (cf. Paragraph 4.2).

As follows, the optimal rate of harvest of a forest is determined by the optimal interdependent harvest ages. The corresponding concept of the stand is the thinning intensity (cf. Section 4.3.1). While negative interdependencies (supply induced price decline, ecological competition, accessibility) tend to reduce the rate of harvest, positive interdependencies (protection, cutting volume) tend to increase the rate of harvest, cf. [4-37]. Differences in the value growth rate might be compensated by differences in the relative land rent and the impact rates such that more stands share equal rotation ages and the rate of harvest increases. Similarly, the rate of harvest increases in the presence of fixed harvest costs (cf. Section 4.4.2.2) which might compensate differences in the net value growth rates, cf. [4-69]. On the other hand, the harvest rate might be restricted (Heaps and Neher 1979). With the dynamics of market exchanges and/ or uncertainty, however, there can be no optimal long-run age class distribution as the rate of harvest is not the ultimate objective.

The general many-aged model [3-38] is an interesting proposal for a unified view on the economics of timber production. Both the Faustmann model [3-2] and the thinning model [3-6] can be considered as special cases of the general uneven-aged management. Depending on the level of observation, trees or stands might be the basic unit of the analysis. In this way, the potential discrepancies between divisible and indivisible capital (Oderwald and Duerr 1990) are dissolved in the model. Indivisible capital is simply the selected unit of the analysis. Whether this is a stand or a tree, or even different parts of a tree are cut at different ages (cf. Li and Löfgren 2000), is freely selectable by the analyst. With negative interdependencies between trees or stands, some trees might be cut prior to the rotation age, or the stands at an earlier age respectively, which corresponds to the conclusions in Oderwald and Duerr (1990). Moreover, the *LEV*s in [5-36] might equally refer to any other net present value of land use, such as agriculture. In this way, the forest model can be interpreted as a general land use model.

6 Summary

When forest owners conduct thinnings in forest stands, their underlying objectives might be various. In the present study, the implications are analyzed when forest owners conduct thinnings in order to generate income with the production of timber. The analysis is restricted to the incentives and opportunities to pursue this aim in an open market of voluntary exchange. In this way, the problem of when to harvest trees prior to the rotation age can be solved within the Faustmann approach as the central action theory of forest economic science.

From the perspective of timber growth, thinnings can be interpreted as subsequent reductions of the initial density in a forest stand. As a consequence, their influences on the timber volume and its structure are qualitatively equivalent. However, due to shorter adjustment periods, lower responsiveness of older trees and potentially less uniform growing areas, the impact of thinnings on the remaining trees is less pronounced. While the stem diameter and the timber volume of the remaining trees are thus expected to increase after thinnings have been conducted, the change of the standing timber volume of the stand depends on the occurrence of density-dependent mortality. Without mortality occurring subsequently, thinnings will decrease the timber volume of the stand.

For the economic analysis, the restriction of the independency and uniformity of the harvest ages of the trees in the Faustmann model is repealed by the introduction of mutual interdependencies between the trees growing in the stand and the opportunity on the part of the forest owner to harvest trees prior to the rotation age. As a consequence, a simultaneous equation system arises which determines the optimal harvest ages of each tree within the constraints of the rule system of a market society. Since any concept of thinning developed in the history of forestry is implemented by the harvest of trees at specific ages prior to the rotation age, the optimal harvest ages of the trees in the stand determine the optimal thinning regime unambiguously.

From the analysis of the implications of the optimal harvest ages of the trees in a forest stand, it follows that thinnings become relevant with heterogeneity concerning the value growth rates of standing trees and competitive pressure concerning the negative impact on the timber value of remaining trees. In order to be sufficient, though, both conditions must outweigh the relative land rent representing the timber harvest opportunities in future rotation periods per standing tree. Whenever thinnings are thus relevant, the reasons to thin a stand more intensively, or less frequently respectively, are given by equal and independent or unequal and interdependent growth. In each case, several trees might share equal optimal harvest ages. In the presence of fixed harvest cost, thinnings are intensified as some of the additional revenues of the remaining trees are not worth the additional cost. With comparatively low variable harvest cost and/ or comparatively small differences in the impacts on the remaining trees, on the other hand, thinnings tend to be conducted from below, while they tend to be applied from above in the opposite case.

The comparative static analysis of the thinning model indicates the direction of the changes in the optimal harvest ages when changes in the investment situation occur. While the optimal rotation age decreases with rising timber prices and falling regeneration costs, the impact on the optimal thinning ages remains ambiguously if the model is not further specified or reduced in its complexity. With the exception of the impact of changes in the rate of interest, the comparative static analysis of the thinning model is thus qualitatively equivalent to the results provided by the Faustmann model.

The analysis indicates various implications. Optimal thinning is basically a two-stage process with a feedback path such that the harvest age of a tree is only optimal given that all other harvest ages are optimal. These interrelations emphasize the relativity of the most profitable trees and, thus, of those trees which promise the greatest increase in intertemporal income. In the same way, forest stands are prevented from being exploited by selective thinnings through the opportunity to regenerate the stand for future timber harvests. With the theoretical basis of an extended Faustmann model with thinnings, additional incentives of forest owners can be analyzed. Accordingly, it can be shown that the relevance of thinnings increases under borrowing constraints and risks of catastrophic events and that thinnings offer the opportunity to adjust the investment in timber growth when unanticipated changes occur.

While the analysis in this study is restricted to thinnings in pure and even-aged forest stands, the underlying problem might be generalized to be applied to different types of forest stands and land-use concepts as well as to entire forests. In each case, the problem of when to harvest simultaneously grown goods (or services) with mutual interdependencies arises. In this sense, differing value growth rates represent mixed forest stands and multiple-use stands, interdependent stands comprised of a single tree represent uneven-aged forest stands and localized and intensive thinnings represent forests. On this account, a uniform approach towards the production of timber, and of land-use via markets in general, is conceivable through simply adjusting the level of observation

7 Appendices

7.1 Appendix 1

The implication of two trees growing in volume at equal rates is analyzed for the impact of the variable harvest cost. If a tree grows in timber volume q^i at a rate α, i.e., if

$$\frac{q_2^i - q_1^i}{q_1^i} := \alpha, \qquad [7\text{-}1]$$

where the subscripts denote different ages with $2 > 1$, the corresponding stem diameter d^i grows at a rate of $\sqrt{(\alpha + 1)w} - 1$ since, cf. [2-4],

$$q_2^i - q_1^i := \frac{\left(d_2^i\right)^2 \pi g_2^i b_2^i}{4} - \frac{\left(d_1^i\right)^2 \pi g_1^i b_1^i}{4} = \alpha \frac{\left(d_1^i\right)^2 \pi g_1^i b_1^i}{4} := \alpha q_1^i$$

$$\Leftrightarrow \quad \left(d_2^i\right)^2 g_2^i b_2^i - \left(d_1^i\right)^2 g_1^i b_1^i = \alpha \left(d_1^i\right)^2 g_1^i b_1^i$$

$$\Leftrightarrow \quad \frac{\left(d_2^i\right)^2 g_2^i b_2^i}{\left(d_1^i\right)^2 g_1^i b_1^i} - 1 = \alpha \qquad [7\text{-}2]$$

$$\Leftrightarrow \quad \frac{d_2^i}{d_1^i} = \sqrt{(\alpha + 1)w} \quad \text{with } w = \frac{g_1^i b_1^i}{g_2^i b_2^i}.$$

If two trees a and b are compared which grow in timber volume at an equal rate α, both diameters are related to each other like the root of ratio of their timber volumes, i.e.

$$\frac{\alpha q^a}{\alpha q^b} = \frac{q^a}{q^b} = \frac{(d^a)^2 \pi g^a b^a}{4} \frac{4}{(d^b)^2 \pi g^b b^b} = \frac{(d^a)^2}{(d^b)^2}$$

$$\Leftrightarrow \quad \sqrt{\frac{q^a}{q^b}} = \frac{d^a}{d^b} := \beta \qquad [7\text{-}3]$$

if $g^a b^a = g^b b^b$ due to the assumption of equal height and shape growth of two competitive trees over a broad range (cf. Section 2.1.3). In the same way, the diameter increments are related since

$$\frac{d_2^a - d_1^a}{d_2^b - d_1^b} = \frac{\sqrt{\alpha + 1}w^a d_1^a - d_1^a}{\sqrt{\alpha + 1}w^b d_1^b - d_1^b} = \frac{d_1^a(\sqrt{\alpha + 1}w^a - 1)}{d_1^b(\sqrt{\alpha + 1}w^b - 1)} = \frac{d_1^a}{d_1^b} = \beta, \qquad [7\text{-}4]$$

employing [7-2] and $g^a b^a = g^b b^b$ again. Accordingly, the diameter increases proportionately to the root of the timber volume ratio. If, for instance, $q^a > q^b$, then the thicker tree a grows in thickness at the factor β compared to the thinner tree.

Though the thicker tree has a higher diameter increment, the unequal change in the variable harvest costs has to be considered. Without price differentials, the net of variable harvest cost unit revenue is decreasing degressively over the diameter, cf. [4-57] in connection with [4-55]. Therefore, equal changes in the diameters result in higher net unit revenue rates of thinner trees. The ratio of the change in the variable harvest costs for the two considered trees, defined as γ, can then be rearranged as, cf. [4-55],

$$\frac{4V}{(d_2^b)^2 \pi l^b b^b} - \frac{4V}{(d_1^b)^2 \pi l^b b^b} = \gamma \left[\frac{4V}{(d_2^a)^2 \pi l^a b^a} - \frac{4V}{(d_1^a)^2 \pi l^a b^a} \right]$$

$$\Leftrightarrow \quad \frac{1}{(d_2^b)^2} - \frac{1}{(d_1^b)^2} = \gamma \left[\frac{1}{(d_2^a)^2} - \frac{1}{(d_1^a)^2} \right]$$

$$\Leftrightarrow \quad \frac{(d_1^b)^2 - (d_2^b)^2}{(d_2^b)^2 (d_1^b)^2} = \gamma \frac{(d_1^a)^2 - (d_2^a)^2}{(d_2^a)^2 (d_1^a)^2}$$

$$\Leftrightarrow \quad \frac{(d_1^b)^2 - \sqrt{\alpha + 1}w^b (d_1^b)^2}{(d_1^a)^2 - \sqrt{\alpha + 1}w^a (d_1^a)^2} = \gamma \frac{(d_2^b)^2 (d_1^b)^2}{(d_2^a)^2 (d_1^a)^2}$$

$$\Leftrightarrow \quad \frac{(d_1^b)^2}{(d_1^a)^2} = \gamma \frac{(d_2^b)^2 (d_1^b)^2}{(d_2^a)^2 (d_1^a)^2}$$

$$\Leftrightarrow \quad \frac{d_2^a}{d_2^b} = \sqrt{\gamma},$$

$$[7\text{-}5]$$

given that $g^a l^a = g^b l^b$ and $l = h$. Since the ratio of the diameters of the two trees is equal to the root of the ratio of their timber volumes by virtue of [7-3], the ratio of the changes in the variable harvest costs is the ratio of the timber volumes according to [7-5]. As $q^a/q^b > \sqrt{q^a/q^b}$ for $q^a > q^b$, the advantage of the higher diameter increment of the thicker tree is overcompensated by the steeper decline of the variable harvest cost of the thinner tree. In this way, the value increment of the thicker tree is less than the value increment of the thinner tree provided both grow in volume at equal rates and price differentials are absent

Acknowledgement

Without the courage of the many authors in the history of science to dare the refutation of their ideas published to be open to criticism, the progression of human knowledge about the observable world would be impossible. In this way, I am indebted to the brilliant thoughts of all writers who took the challenge of the scientific venture without which the present study would have been unfeasible. In order to acknowledge these contributions, I tried to refer to the original sources of the ideas whenever possible and meaningful. Nevertheless, only the smallest subset of contributors could be cited.

From all these teachers, the most inspiring to me has been Professor Peter Deegen. It was his ideas and his view of science which motivated me to engage in scientific writing. Not only has he enabled the material requirements for this study, but he has enabled the freedom of thought which promised to be the best investment into scientific progress. For the countless valuable suggestions and discussions of an astonishing versatility, I am deeply grateful.

For the professional and overall friendly reception into his team, I would like to thank Professor Norbert Weber. It was very inspiring to share his ideas on problems which offered a different and thus highly valuable perspective to me. Many thanks for these views beyond the border of forestry economics.

For their offer to prepare a review of this study and for the helpful comments, I am very grateful to Professor Sun Joseph Chang and Professor Heinz Röhle. Furthermore, I would like to thank Professor Sven Wagner and Andreas Halbritter for the helpful discussions which improved the quality of this study markedly. For the helpful and friendly atmosphere during the preparation of this study, I would like to thank all members of the Institute of Forest Economics and Forest Management Planning.

It is acknowledged that parts of this study (esp. parts of sections 3.1, 4.2.1, 4.5 and 5.2.3) are reprinted from an earlier publication from Forest Science published by the Society of American Foresters (cf. Coordes 2014).

References

Altherr, E. (1966). Die Bedeutung des Pflanzverbandes für die Leistung der Fichtenbestände. *Allgemeine Forst Zeitschrift*, 21: 191-200.

Amacher, G. S., Koskela, E. and Ollikainen, M. (2008). Deforestation and land use under insecure property rights. *Environment and Development Economics*, 14: 281-303.

Amacher, G. S., Ollikainen, M. and Koskela, E. (2009). *Economics of Forest Resources*. The MIT Press, Cambridge.

Anderson, F. J. (1976). Control Theory and the Optimum Timber Rotation. *Forest Science*, 22: 242-246.

Assmann, E. (1970). *The principles of forest yield study*. Pergamon Press, Oxford.

Begon, M., Harper, J. L. and Townsend, C. R. (1990). *Ecology. Individuals, Populations and Communities*. 2nd Edition. Blackwell Scentific Publications, Boston.

Bellman, R. (1957). *Dynamic Programming*. Princeton University Press, Princeton.

Bendz-Hellgren, M. and Stenlid, J. (1998). Effects of clear-cutting, thinning, and wood moisture content on the susceptibility of Norway spruce stumps to Heterobasidion annosum. *Canadian Journal of Forest Research*, 28: 759-765.

Betters, D. R., Steinkamp, E. A. and Turner, M. T. (1991). Singular Path Solutions and Optimal Rates of Thinning Even-Aged Forest Stands. *Forest Science*, 37: 1632-1640.

Beverton, R. J. and Holt, S. J. (1957). *On the Dynamics of Exploited Fish Populations*. Chapman and Hall, London.

Binkley, C. S. (1987). When Is the Optimal Economic Rotation Longer than the Rotation of Maximum Sustained Yield? *Journal of Environmental Economics and Management*, 14: 152-158.

Binkley, C. S. (1993). Long-run Timber Supply: Price Elasticity, Inventory Elasticity, and the Use of Capital in Timber Production. *Natural Resource Modeling*, 7: 163-181.

Borchert, H. (2002). The Economically Optimal Amount of Timber Cut in Forests - An Approach by Control Theory. *Schriften zur Forstökonomie* Bd. 24, J. D. Sauerländers, Frankfurt a. M.

Boyden, S., Binkley, D. and Stape, J. L. (2008). Competition Among Eucalyptus Trees Depending on Genetic Variation and Resource Supply. *Ecology*, 89: 2850-2859.

Bradford, J. B. and Palik, B. J. (2009). A comparision of thinning methods in red pine: consequences for the stand-level growth and tree diameter. *Canadian Journal of Forest Research*, 39: 489-496.

Brazee, R. J. and Bulte, E. (2000). Optimal Harvesting and Thinning with Stochastic Prices. *Forest Science*, 46: 23-31.

Brodie, J. D., Adams, D. M. and Kao, C. (1978). Analysis of Economic Impacts on Thinning and Rotation for Douglas-fir, Using Dynamic Programming. *Forest Science*, 74: 513-523.

Brown, G. S. (1965). Point density in stems per acre. *New Zealand Forestry Service Research Notes*, 38: 1-11.

Buchanan, J. M. (1964). What Should Economists Do? *Southern Economic Journal*, 30: 213-222.

Buchanan, J. M. (1999). *The Demand and Supply of Public Goods*. The Collected Works of James M. Buchanan, Vol. 5, Liberty Fund, Indianapolis.

Buchanan, J. M. and Brennan, G. (2000). *The Reason of Rules: Constitutional Political Economy*. Collected Works of James M. Buchanan, Vol. 10, Liberty Fund, Indianapolis.

Busse, J. and Jaehn, R. (1925). Wuchsraum und Zuwachs (Wermsdorfer Fichten-Kulturversuch). *Mitteilungen der sächsisch forstlichen Versuchsanstalt zu Tharandt*, II, 6.

Cameron, A. D. (2002). Importance of early selective thinning in the development of long-term stand stability and improved log quality: a review. *Forestry*, 75: 25-35.

Campoe, O. C., Stape, J. L., Nouvellon, Y., Laclau, J.-P., Bauerle, W. L., Binkley, D. and Maire, G. L. (2013). Stem production, light absorption and light use efficiency between dominant and non-dominant trees of Eucalyptus grandis across a productivity gradient in Brazil. *Forest Ecology and Management*, 288: 14-20.

Cao, T., Hyytiäinen, K., Tahvonen, O. and Valsta, L. (2006). Effects of initial stand states on optimal thinning regime and rotation of Picea abies stands. *Scandinanvian Journal of Forest Research*, 21: 388-398.

Carlowitz, H. C. (1713). *Sylvicultura Oeconomica. Haußwirtschaftliche Nachricht und Naturgemäßige Anweisung Zur Wilden Baum-Zucht.* J.F. Braun, Leipzig.

Cawrse, D. C., Better, D. R. and Kent, B. M. (1984). A Variational Soltuion Technique for Determining Optimal Thinning and Rotation Schedules. *Forest Science*, 30: 793-802.

Chang, S. J. (1981). Determination of the Optimal Growing Stock and Cutting Cycle for an Uneven-Aged Stand. *Forest Science*, 27: 739-744.

Chang, S. J. (1983). Rotation age, management intensity, and the economic factors of timber production: do changes in stumpage price, interest rate, regeneration cost, and forest taxation matter? *Forest Science*, 29: 267–277.

Chang, S. J. (1984). A simple production function model for variable density growth and yield modeling. *Canadian Journal of Forest Research*, 14: 783-788.

Chang, S. J. (1998). A generalized Faustmann model for the determination of optimal harvest age. *Canadian Journal of Forest Research*, 28: 652-659.

Chang, S. J. (2001). One formula, myriad applications - 150 years of practicing the Faustmann Formula in Central Europe and the USA. *Forest Policy and Economics*, 2: 97-99.

Chang, S. J. and Deegen, P. (2011). Pressler's indicator rate formula as a guide for forest management. *Journal of Forest Economics*, 17: 258-266.

Clark, C. W. (2005). *Mathematical Bioeconomics.* Optimal Management of Renewable Resources. 2nd Edition, Wiley & Sons, Inc., New Jersey.

Clark, C. W. and De Pree, J. D. (1979). A Simple Linear Model for the Optimal Exploitation of Renewable Resources. *Applied Mathematics and Optimization*, 5: 181-196.

Clark, C. W. and Munro, G. R. (1975). The Economics of Fishing and Modern Capital Theory: A Simplified Approach. *Journal of Environmental Economics and Management*, 2: 92-106.

Clason, T. R. (1994). Impact of intraspecific competition on growth and financial development of loblolly pine plantations. *New Forests*, 8: 185-210.

Conrad, J. M. (2010). *Resource Economics*. 2nd Edition, Cambridge University Press, Cambridge.

Conrad, J. M. and Clark, C. W. (1987). *Natural Resource Economics*. Cambridge University Press, Cambridge.

Coordes, R. (2013). Influence of planting density and rotation age on the profitability of timber production for Norway spruce in Central Europe. *European Journal of Forest Research*, 132: 297-311.

Coordes, R. (2014). Thinnings as unequal harvest ages in even-aged stands. *Forest Sience*, in press, doi:10.5849/forsci.13-004.

Cotta, H. (1817). *Anweisung zum Waldbau*. 2nd Edition, Arnoldische Buchhandlung, Dresden.

Deegen, P. (2001). *Aufforstung und Holzeinschlag als Investitionsprobleme in einer statischen Welt*. Habilitation thesis, TU Dresden, Faculty of Forest, Geo and Hydro Sciences, Dresden.

Deegen, P. (2002). Zur Relevanz von Modellen der intertemporalen Ressourcenökonomie für die finanzielle Analyse naturnaher Waldformen. *Forst und Holz*, 57: 654-659.

Deegen, P. (2012). *Economics of the External and Extended Orders of Markets and Politics and their Application in Forestry*. Paper submitted at the international IUFRO conference "New Frontiers of Forest Economics", Chaired by Prof. Shashi Kant, ETH Zürich, Switzerland.

Deegen, P., Hostettler, M. and Navarro, G. A. (2011). The Faustmann model as a model for a forestry of prices. *European Journal of Forest Research*, 130, 353-368.

Deleuze, C. and Houllier, F. (1995). Prediction of stem profile of Picea abies using a process-based tree. *Tree Physiology*, 15: 113-120.

Drew, T. J. and Flewelling, J. W. (1977). Some recent Japanese theories of yield density relationships and their application to monterey pine plantations. *Forest Science*, 23: 517–534.

Duerr, W. A. (1960). *Fundamentals of Forestry Economics*. McGraw-Hill Book Company, New York.

Duerr, W. A. (1993). *Introduction to Forest Resource Economics*. International Edition. McGraw-Hill, Inc., New York.

Duerr, W. A. and Duerr, J. B. (1975). The Role of Faith in Forest Resource Management. In: F. Rumsey and W. Duerr (Hrsg.): *Social Sciences in Forestry*, pp. 30-41, W.B. Saundersm Philadelphia.

Eckmüller, O. and Sterba, H. (2000). Crown condition, needle mass, and sapwood area relationships of Norway spruce (Picea abies). *Canadian Journal of Forest Research*, 30: 1646-1654.

Faustmann, M. (1849). Berechnung des Wertes welchen Waldboden sowie noch nicht haubare Holzbestände für die Waldwirtschaft besitzen. *Allgemeine Forst- und Jagdzeitung*, 15, 441–455.

Fisher, I. (1930). *The Theory of Interest*. Porcupine Press, Philadelphia.

Gaffney, M. (2008). Keeping Land in Capital Theory: Ricardo, Faustmann, Wicksell, and George. *American Journal of Economics and Sociology*, 67: 119–141.

Garcia-Gonzalo, J., Peltola, H., Briceño-elizondo, E. and Kellomäki, S. (2007). Changed thinning regimes may increase carbon stock under climate change: A case study from a Finnish boreal forest. *Climatic Change*, 81: 431-454.

Gardiner, B. A. and Quine, C. P. (2000). Management of forests to reduce the risk of abiotic damage - a review with particular reference to the effects of strong winds. *Forest Ecology and Management*, 135: 261-277.

Getz, W. M. and Haight, R. G. (1989). *Population Harvesting: Demographic Models of Fish, Forest, and Animal Resources*. Princeton Univerity Press, Princeton.

Gizachew, B., Brunner, A. and Øyen, B.-H. (2012). Stand responses to initial spacing in Norway spruce. *Scandinavian Journal of Forest Research*, 27: 637-648.

Gong, P. and Löfgren, K.-G. (2010). Did Presslerfully understand how to use the indicator per cent? *Journal of Forest Economics*, 16: 195-203.

Gspaltl, M., Bauerle, W., Binkley, D. and Sterba, H. (2013). Leaf area and light use efficiency patterns of Norway spruce under different thinning regimes and age classes. *Forest Ecology and Management*, 288: 49-59.

Haight, R. G. (1985). A Comparison of Dynamic and Static Economic Models of Uneven-Aged Stand Management. *Forest Science*, 31: 957-974.

Haight, R. G. (1987). Evaluating the Efficiency of Even-Aged and Uneven-Aged Stand Management. *Forest Science*, 33: 116-134.

Halbritter, A. and Deegen, P. (2011). Economic analysis of exploitation and regeneration in plantations with problematic site productivity. *Journal of Forest Economics*, 17: 319– 334.

Hall, D. O. (1983). Financial Maturity for Even-Aged and All-Aged Stands. *Forest Science*, 29: 833-836.

Hartig, G. L. (1791). *Anweisung zur Holzzucht für Förster.* Neue Akademische Buchhandlung, Marburg.

Hartman, R. (1976). The Harvesting Decision When a Standing Forest Has Value. *Economic Inquiry*, 14: 52-58.

Hayek, F. A. von (1942). Scientism and the Study of Society. Part I. *Economica*, 35: 267-291.

Hayek, F. A. von (1945). The Use of Knowledge in Society. *The American Economic Review*, 35: 519-530.

Hayek, F. A. von (1964). Kinds of Order in Society. *New Individualist Review*, 3: 3-12.

Hayek, F. A. von (1964). The Theory of Complex Phenomena. In: M. Bunge (Hrsg.): *The Critical Approach to Science and Philosophy. Essays in Honor of K.R. Popper*, pp. 332-349, The Free Press of Glencoe, New York and London.

Hayek, F. A. von (1973). *Law, Legislation and Liberty, Vol. 1: Rules and Order.* The University of Chicago Press, Chicago.

Hayek, F. A. von (2002). Competition as a Discovery Procedure. *The Quarterly Journal of Austrian Economics*, 5: 9-23.

Heaps, T. and Neher, P. A. (1979). The Economics of Forestry when the Rate of Harvest is Constrained. *Journal of Environmental Economics and Management*, 6: 297-319.

Heyer, C. (1854). *Der Waldbau oder die Forstproductenzucht.* B. Teubner, Leipzig.

Hirshleifer, J. (1970). *Investment, Interest and Capital.* Prentice-Hall, Upper Saddle River.

Hirshleifer, J., Glazer, A. and Hirshleifer, D. (2005). *Price Theory and Applications.* 7th Edition, Cambridge University Press, Cambridge.

Homann, K. and Suchanek, A. (2005). *Ökonomik. Eine Einführung.* 2nd Edition, Mohr Siebeck, Tübingen.

Huber, B. (1928). Weitere quantitative Untersuchungen über das Wasserleitungssystem der Pflanzen. *Jahrbücher für wissenschaftliche Botanik*, 67: 877-959.

Hundeshagen, J. C. (1821). *Encyclopädie der Forstwissenschaft.* H. Laupp, Tübingen.

Hyde, W. F. (1980). *Timber Supply, Land Allocation, and Economic Efficiency.* The John Hopkins University Press, Baltimore, Maryland.

Hyde, W. F. (2012). *The Global Economics of Forestry.* RFF Press, New York.

Hyytiäinen, K. and Tahvonnen, O. (2002). Economics of Forest Thinnings and Rotation Periods for Finnish Conifer Cultures. *Scandinavian Journal for Forest Research*, 17: 274-288.

Hyytiäinen, K., Tahvonen, O. and Valsta, L. (2005). Optimum Juvenile Density, Harvesting, and Stand Structure in Even-Aged Scots Pine Stands. *Forest Science*, 51: 120-133.

Inoue, A., Miyake, M. and Nishizono, T. (2004). Allometric model of the Reineke equation for Japanese cypress (Chamaecyparis obtusa) and red pine (Pinus densiflora) stands. *Journal of Forest Research*, 9: 319-324.

Johansson, P.-O. and Löfgren, K.-G. (1985). *Economics of Forestry and Natural Resources.* Basil Blackwell, Oxford, New York.

Kahneman, D. (2011). *Thinking, Fast and Slow.* Penguin Books London.

Kalies, E. L., Chambers, C. L. and Covington, W. W. (2010). Wildlife responses to thinning and burning treatments in southwestern conifer forests: A meta-analysis. *Forest Ecology and Management*, 259: 333-342.

Kaufmann, M. R. and Troendle, C. A. (1981). The Relationship of Leaf Area and Foliage Biomass to Sapwood Conducting Area in Four Subalpine Forest Tree Species. *Forest Science*, 27: 477-482.

Kilkki, P. and Väisänen, U. (1969). Determination of the Optimum Cutting Policy for the Forest Stand by Means of Dynamic Programming. *Acta Forestalia Fennica*, 102: 5-22.

Kimmins, J. P. (1987). *Forest ecology.* Macmillan Publishing Company, New York.

Klemperer, D. W. (1996). *Forest Resource Economics and Finance.* McGraw-Hill, Inc., New York.

Knight, F. H. (1921). *Risk, Uncertainty and Profit.* Houghton Mifflin Company, New York.

Koskela, E. and Ollikainen, M. (2001). Optimal Private and Public Harvesting under Spatial and Temporal Interdependence. *Forest Science*, 47: 484–496.

Kraft, G. (1884). *Beiträge zur Lehre von den Durchforstungen, Schlagstellungen und Lichtungshieben*. Kindworth's Verlag, Hannover.

Kramer, H. and Spellmann, H. (1980). Beiträge zur Bestandesbegründung der Fichte. *Schriften aus der Forstlichen Fakultät der Universität Göttingen und der Niedersächsischen Versuchsanstalten*, Band 64, J. D. Sauerländer's Verlag, Frankfurt a. M.

Lässig, R. (1991). *Zum Wachstum von Fichtensolitären* [Picea abies (L.) Karst.] in Südwestdeutschland. Doctoral thesis, Forestry Faculty, University Freiburg.

Li, C.-Z. and Löfgren, K.-G. (2000). A Theory of Red Pine (Pinus koraiensis) Management for Both Timber and Commercial Seeds. *Forest Science*, 46: 284-290.

Löfgren, K.-G. (1985). Effect on the Socially Optimal Rotation Period in Forestry of Biotechnological Improvements of the Growth Function. *Forest Ecology and Management*, 10: 233-249.

Lu, H.-C. and Chang, S. J. (1996). The impact of declining forest productivity on the optimal rotation age of a timber stand. In: *Proceedings of the 25th Annual Southern Forest Economics Workshop*, April 1995, pp. 281–292, LA. Mississippi State University, New Orleans.

Machlup, F. (1958). Equilibrium and Disequilibrium: Misplaced Concreteness and Disguised Politics. *The Economic Journal*, 68: 1-24.

Machlup, F. (1959). Statics and Dynamics: Kaleidoscopic Words. *Southern Economic Journal*, 26: 91-110.

Magin, R. (1952). Zuwachsleistungen der soziologischen Baumklassen in langfristig beobachteten Versuchsflächen. *Forstwissenschaftliches Centralblatt*, 71: 225-243.

Mäkelä, A. (2002). Derivation of stem taper from the pipe theory in a carbon balance. *Tree Physiology*, 22: 891–905.

Mäkinen, H. and Hein, S. (2006). Effect of wide spacing on increment and branch properties. *European Journal of Forest Research*, 125: 239 – 248.

Mäkinen, H. and Isomäki, A. (2004). Thinning intensity and growth of Norway spruce stands in Finland. *Forestry*, 77: 349-364.

Marshall, A. (1922). *Principles of Economics: An Introductory Analysis*. 8th Edition, Macmillan, London.

McConnell, K. E., Daberkow, J. N. and Hardie, I. W. (1983). Planning Timber Production with Evolving Prices and Costs. *Land Economics*, 59: 292-299.

Medema, E. L. and Lyon, G. W. (1985). The Determination of Financial Rotation Ages for Coppicing Tree Species. *Forest Science*, 31: 398-404.

Mohler, C. L., Marks, P. L. and Sprugel, D. G. (1978). Stand Structure and Allometry of Trees During Self-Thinning of Pure Stands. *Journal of Ecology*, 66: 599-614.

Moser, W. G. (1757). *Grundsaetze der Forst-Oeconomie*. Heinrich Ludwig Brönner, Frankfurt und Leipzig.

Musgrave, A. (1981). 'Unreal Assumptions' in Economic Therory: The F-Twist Untwisted. *Kyklos*, 34: 377-387.

Näslund, B. (1969). Optimal Rotation and Thinning. *Forest Science*, 15: 446-451.

Neher, P. A. (1990). *Natural Resource Economics - Conservation and Exploitation*. Cambridge University Press, Cambridge.

Neill, A. R. and Puettmann, K. J. (2013). Managing for adaptive capacity: thinning improves food availability for wildlife and insect pollinators under climate change conditions. *Canadian Journal of Forest Research*, 43: 428-440.

Newman, D. H., Gilbert, C. B. and Hyde, W. F. (1985). The Optimal Forest Rotation with Evolving Prices. *Land Economics*, 61: 213-218.

Nultsch, W. (2001). *Allgemeine Botanik*. Georg Thieme Verlag, Stuttgart.

Nyland, R. D. (2002). *Silviculture: Concepts and Applications*. 2nd Edition, McGraw-Hill, New York.

Oderwald, R. G. and Duerr, W. A. (1990). König-Faustmannism: A Critique. *Forest Science*, 36: 169-174.

O'Hara, K. L. and Oliver, C. D. (1988). Three-Dimensional Representation of Douglas-Fir Volume Growth: Comparision of Growth and Yield Models with Stand Data. *Forest Science*, 34: 724-743.

Oliver, C. D. and Larson, B. C. (1996). *Forest Stand Dynamics*. John Wiley and Sons, New York.

Petersen, R. and Spellmann, H. (1993). Fichtenverbandsversuch Braunlage 51. *Forst und Holz*, 48: 83 – 86.

Pfeiffer, J. F. (1781). *Grundriss der Forstwissenschaft. Zum Gebrauch dirigierender Forst- und Kameralbedienten, auch Privatguthsbesitzern.* E.F. Schwan, Mannheim.

Pfeil, W. (1820). *Vollständige Anleitung zur Behandlung, Benutzung und Schätzung der Forsten.* Erster Band. Darnmansche Buchhandlungen, Züllichau und Freistadt.

Popper, K. R. (1966). *The Open Society and its Enemies.* Princeton University Press, Princeton.

Popper, K. R. (1983). The rationality principle. In: D. Miller (Hrsg.): *A Pocket Popper*, pp. 357-378. Fontana, Oxford.

Popper, K. R. (2002a). *Conjectures and Refutations.* Routledge, London.

Popper, K. R. (2002b). *The Logic of Scientific Discovery.* Routledge, London.

Popper, K. R. (2010). *Lesebuch.* 2nd Edition, Mohr Siebeck, Tübingen.

Powers, M. D., Palik, B. J., Bradford, J. B., Fraver, S. and Webster, C. R. (2010). Thinning method and intensity influence long-term mortality trends in a red pine forest. *Forest Ecology and Management*, 260: 1138–1148.

Pressler, M. R. (1860). Zur Verständigung über den Reinertragswaldbau und dessen Betriebsideal. Zweiter Artikel. Aus der Holzzuwachlehre. *Allgemeine Forst- und Jagd-Zeitung*, 36: 173-191. Translated by: Löwenstein, W. and Wirkner, J. R. (1995). For the comprehension of net revenue silviculture and the management objectives derived thereof. *Journal of Forest Economics*, 1: 45-87.

Pressler, M. R. (1865). *Das Gesetz der Stammbildung.* Arnoldische Buchhandlung, Leipzig.

Pressler, M. R. (1865). *Der rationelle Forstwirth und dessen Reinertrags-Forstwirtschaft inner und außer dem Walde.* Flugblatt No. 1. Die Forstwirthschaft der sieben Thesen. Woldemar Türk, Dresden.

Pretzsch, H. (2000). Die Regeln von Reineke, Yoda und das Gesetz der räumlichen Allometrie. *Allg. Forst- und Jagdzeitung*, 171: 205-210.

Pretzsch, H. (2005). Stand density and growth of Norway spruce (Picea abies (L.) Karst.) and European beech (Fagus sylvatica L.): evidence from long-term experimental plots. *European Journal of Forest Research*, 124: 193-205.

Pretzsch, H. (2009). Forest Dynamics, Growth and Yield. Springer Verlag, Berlin, Heidelberg.

Pretzsch, H. and Biber, P. (2005). A Re-Evaluation of Reineke's Rule and Stand Density Index. *Forest Science*, 51: 304-320.

Puettmann, K. J., Coates, K. D. and Messier, C. (2009). *A Critique of Silviculture: Managing for Complexity*. Island Press, Washington.

Pukkala, T. and Miina, J. (1998). Tree-selection algorithms for optimizing thinning using a distance-dependent growth model. *Canadian Journal of Forest Research*, 28: 693-702.

Pukkala, T., Miina, J., Kurttila, M. and Kolström, T. (1998). A Spatial Yield Model for Optimizing the Thinning Regime of Mixed Stands of Pinus sylvestris and Picea abies. *Scandinavian Journal of Forest Research*, 13: 31-42.

Raup, H. M. (1966). The View from John Sanderson's Farm: A Perspective for the Use of the Land. *Forest History*, 10: 2-11.

Reed, W. J. (1984). The Effect of the Risk of Fire on the Optimal Rotation of a Forest. *Journal of Environmental Economics and Management*, 11: 180-190.

Reed, W. J. (1987). Protecting a forest against fire: Optimal protection patterns and harvest policies. *Natural Resource Modeling*, 2: 23-54.

Reed, W. J. and Apaloo, J. (1991). Evaluating the effects of risk on the economics of juvenile spacing and commercial thinning. *Canadian Journal of Forest Research*, 21: 1390-1400.

Reineke, L. H. (1933). Perfecting a Stand-Density Index for Even-Aged Forests. *Journal of Agricultural Research*, 46: 627-633.

Roise, J. P. (1986). An Approach for Optimizing Residual Diameter Class Distributions When Thinning Even-Aged Stands. *Forest Science*, 32: 871-881.

Rydberg, D. and Falck, J. (1998). Designing the urban forest of tomorrow: Pre-commercial thinning adapted for use in urban areas in Sweden. *Arboricultural Journal*, 22: 147-171.

Salo, S. and Tahvonen, O. (2002). On the Optimality of a Normal Forest with Multiple Land Classes. *Forest Science*, 48: 530–542.

Salo, S. and Tahvonen, O. (2004). Renewable resource with endogenous age classes and allocation of land. *American Journal of Agricultural Economics*, 86: 513-530.

Samuelson, P. A. (1976). Economics of Forestry in an Evolving Society. *Economic Inquiry*, 14: 466-492.

Samuelson, P. A. (1983). *Foundations of Economic Analysis*. Enlarged Edition, Harvard Univerity Press, Cambridge.

Schreuder, G. F. (1971). The Simultaneous Determination of Optimal Thinning Schedule and Rotation for an Even-Aged Forest. *Forest Science*, 17: 333-339.

Schwappach, A. (1911). *Die Rotbuche*. Wirtschaftliche und statische Untersuchungen der forstlichen Abteilung der Hauptstation des forstlichen Versuchswesens in Eberswalde. Neudamm, Eberswalde.

Schwinning, S. and Weiner, J. (1998). Mechanisms determining the degree of size asymmetry in competition among plants. *Oecologia*, 113: 447-455.

Scott, W., Meade, R., Leon, R., Hyink, D. and Miller, R. (1998). Planting density and tree-size relations in coast Douglas-fir. *Canadian Journal of Forest Research*, 28: 74-78.

Serengil, Y., Gökbulak, F., Özhan, S., Hizal, A., Sengönül, K., Balci, A. N. and Özyuvaci, N. (2007). Hydrological impacts of a slight thinning treatment in a deciduous forest ecosystem in Turkey. *Journal of Hydrology*,333: 569–577.

Shinozaki, K., Yoda, K., Hozumi, K. and Kira, T. (1964a). A Quantitative Analysis of Plant Form - The Pipe Model Theorie. I. Basic Analyses. *Japanese Journal of Ecology*, 14: 97-105.

Shinozaki, K., Yoda, K., Hozumi, K. and Kira, T. (1964b). A Quantitative Analysis of Plant Form - The Pipe Model Theory. II. Further Evidence of the Theory and its Application in Forest Ecology. *Japanese Journal of Ecology*, 14: 133-139.

Smith, D. M., Larson, B. C., Kelty, M. J. and Ashton, P. M. (1997). *The Practice of Silviculture*. 9th Edition, Jon Wiley & Sons, Inc., New York

Smith, V. L. (1991). *Papers in Experimental Economics*. Cambridge University Press, Cambridge.

Solberg, B. and Haight, R. G. (1991). Analysis of Optimal Economic Management Regimes for Picea abies Stands Using a Stage-Structured Optimal-Control Model. *Scandinavian Journal of Forest Research*, 6: 559-572.

Strand, H. (1969). Economic Analysis of a Basis of Land-use Policy. In: A. Svendsrud (Hrsg.), *Readings in forest economics*. pp. 241-249, Universiteteforlaget, Oslo, Bergen, Tromso.

Strang, W. J. (1983). On the Optimal Harvesting Decision. *Economic Inquiry*, 21: 576-583.

Swallow, S. K. and Wear, D. N. (1993). Spatial Interactions in Multiple-Use Forestry and Substitution and Wealth Effects for the Single Stand. *Journal of Environmental Economics and Management*, 25: 103-120.

Swallow, S. K., Parks, P. J. and Wear, D. N. (1990). Policy-Relevant Nonconvexities in the Production of Multiple Forest Benefits? *Journal of Environmental Economics and Management*, 19: 264-280.

Tahvonen, O. (2009). Optimal choice between even- and uneven-aged forestry. *Natural Resource Modelling*, 22: 289-321.

Tahvonen, O. and Salo, S. (1999). Optimal Forest Rotation with in Situ Preferences. *Journal of Environmental Economics and Management*, 37: 106-128.

Tahvonen, O. and Viitala, E.-J. (2006). Does Faustmann Rotation Apply to Fully Regulated Forests? *Forest Science*, 52: 23-30.

Tahvonen, O., Pukkala, T., Laiho, O., Lähde, E. and Niinimäki, S. (2010). Optimal management of uneven-aged Norway spruce stands. *Forest Ecology and Management*, 260: 106-115.

Tahvonen, O., Salo, S. and Kuuluvainen, J. (2001). Optimal forest rotation and land values under a borrowing constraint. *Journal of Economic Dynamics and Control*, 25: 1595-1627.

Tait, D. E. (1986). A dynamic programming solution of financial rotation ages for coppicing tree spcies. *Canadian Journal of Forest Research*, 16: 799-801.

Tang, J., Qi, Y., Xu, M., Mission, L. and Goldstein, A. H. (2005). Forest thinning and soil respiration in a ponderosa pine plantation in the Sierra Nevada. *Tree Physiology*, 25: 57-66.

Tang, S., Meng, C. H., Meng, F.-R. and Wang, Y. H. (1994). A growth and self-thinning model for pure even-age stands: theory and applications. *Forest Ecology and Management*, 70: 67-73.

Thünen, J. H. von (1842). *Der isolirte Staat in Beziehung auf Landwirthschaft und Nationalökonomie. 2., verm. u. verb. Aufl. Teil 1: Untersuchungen über den Einfluß, den die Getreidepreise, der Reichthum des Bodens und die Abgaben auf den Ackerbau ausüben.* Leopold, Rostock. Translated by: Wartenberg, C. M. (1966). *The Isolated State.* Pergamon Press, Oxford and New York.

Thünen, J. H. von (1875). *Der isolierte Staat in Beziehung auf Landwirthschaft und Nationalökonomie. Dritter Theil. Grundsätze zur Bestimmung der Bodenrente, der vortheilhaftesten Umtriebszeit und des Werths der Holzbestände von verschiedenem Alter für Kieferwaldungen.* 3rd Edition, Verlag von Wiegandt, Hempel und Parev, Berlin. Translated by: Suntum, U. van (2009). *The Isolated State in Relation to Agriculture and Political Economy, Part III : Principles for the Determination of Rent, the Most Advantageous Rotation Period and the Value of Stands of Varying Age in Pinewoods.* Palgrave Macmillan Ltd., New York.

Valentine, H. T. (1985). Tree-growth Models: Derivations Employing the Pipe-model Theory. *Journal of Theoretical Biology*, 117: 579-585.

Valentine, H. T. (1988). A Carbon-balance Model of Stand Growth: a Derivation Employing Pipe-model Theory and the Self-thinning Rule. *Annals of Botany*, 62: 389-396.

Vanclay, J. K. (1994). *Modelling Forest Growth and Yield.* CAB International, Wallingford.

Vanselow, K. (1950). Einfluß des Pflanzverbandes auf die Entwicklung reiner Fichtenbestände. II. *Forstwissenschaftliches Centralblatt*, 69: 497-527.

Vanselow, K. (1956). Einfluss des Pflanzverbandes auf die Entwicklung reiner Fichtenbestände III. *Forstwissenschaftliches Centralblatt*, 75: 193-207.

Varian, H. R. (2010). *Intermediate Microeconomics.* 8th Edition, W. W. Norton & Company, New York.

Vertessy, R. A., Benyon, R. G., O'Sullivan, S. K. and Gribben, P. R. (1995). Relationships between stem diameter, sapwood area, leaf area and transpiration in a young mountain ash forest. *Tree Physiology*, 15: 559-567.

Vettenranta, J. and Miina, J. (1999). Optimizing Thinnings and Rotation of Scots Pine and Norway Spruce Mixtures. *Silva Fennica*, 33: 73-84.

Vincent, J. R. and Binkley, C. S. (1993). Efficient Multiple-Use Forestry May Require Land-Use Specialization. *Land Economics*, 69: 370-376.

Wacker, H. and Blank, J. E. (1998). *Ressourcenökonomik I: Einführung in die Theorie regenerativer natürlicher Ressourcen.* R. Oldenbourg Verlag, München, Wien.

Waring, R. H., Schroeder, P. E. and Oren, R. (1982). Applications of the pipe model theory to predict canopy leaf area. *Canadian Journal of Forest Research*, 12: 556-560.

Weiskittel, A. R., Hann, D. W., Kershaw Jr., J. A. and Vanclay, J. K. (2011). *Forest Growth and Yield Modeling*. Wiley-Blackwell, Chichester.

Weller, D. E. (1987). A Reevaluation of the -3/2 Power Rule of Plant Self-Thinning. *Ecological Monographs*, 57: 23-43.

Weng, S.-H., Kuo, S.-R., Guan, B. T., Chang, T.-Y., Hsu, H.-W. and Shen, C.-W. (2007). Microclimatic responses to different thinning intensities in a Japanese cedar plantation of northern Taiwan. *Forest Ecology and Management*, 241: 91-100.

West, P. W. (2009). *Tree and Forest Measurement*. Springer-Verlag, Berlin.

White, J. (1981). The Allometric Interpretation of the Self-thinning Rule. *Journal of Theoretical Biology*, 89: 475-500.

Wilson, D. S. and Puettmann, K. J. (2007). Density management and biodiversity in young Douglas-fir forests: Challenges of managing across scales. *Forest Ecology and Management*, 246: 123–134.

Withehead, D., Edwards, W. R. and Jarvis, P. G. (1984). Conducting sapwood area, foliage area, and permeability in mature trees of Piceasitchensis and Pinuscontorta. *Canadian Journal of Forest Research*, 14: 940-947.

Yin, R. and Newman, D. H. (1995). Optimal Timber Rotations with Evolving Prices and Costs Revisited. *Forest Science*, 41: 477-490.

Yoda, K., Kira, T., Ogawa, H. and Hozumi, K. (1963). Self-thinning in overcrowded pure stands under cultivated and natural conditions. *Journal of the Institute of Polytechnics*, 14: 107-129.

Zeide, B. (1985). Tolerance and Self-Tolerance of Trees. *Forest Ecology and Management*, 13: 149-166.

Zeide, B. (1995). A relationship between size of trees and their number. *Forest Ecology and Management*, 72: 256-272.

Zeide, B. (2001). Thinning and Growth: A Full Turnaround. *Journal of Forestry*, 99: 20-25.

Zeide, B. (2004). Intrinsic units in growth modeling. *Ecological Modelling*, 175: 249–259.

Zeide, B. (2005). How to measure density. *Trees*, 19: 1-14.

Zhang, D. and Pearse, P. H. (2011). *Forest Economics*. UBC Press, Vancouver.

Zhang, S. Y., Chauret, G., Swift, D. E. and Duchesne, I. (2006). Effects of precommercial thinning on tree growth and lumber quality in a jack pine stand in New Brunswick, Canada. *Canadian Journal of Forest Research*, 36: 945-952.

Zhang, S., Burkhart, H. E. and Amateis, R. L. (1996). Modeling individual tree growth for juvenile loblolly pine plantations. *Forest Ecology and Management*, 89: 157-172.

Index